The Water, Energy and Food Security Nexus

It is becoming increasingly recognized that for the optimal sustainable development and use of natural resources, an integrated approach to water management, agriculture, food security and energy is required. This 'nexus' is now the focus of major attention by researchers, policymakers and practitioners.

In this book, the authors show how these issues are being addressed in India as part of its economic development, and how these examples can provide lessons for other developing nations. They address the conflicting claims of water resources for irrigation and hydropower, where both are scarce at the national level for fostering water and energy security. They also consider the relationship between water for irrigated agriculture and household use and its impact on rural poverty. They identify weaknesses in the current hydropower development program in India that are preventing it from being an ecologically sustainable, socially just and economically viable solution to meeting growing energy demand.

The empirical analyses presented show the enormous scope for co-management of water, energy, agricultural growth and food security through appropriate technological interventions and market instruments.

M. Dinesh Kumar is Executive Director of the Institute for Resource Analysis and Policy (IRAP), and is based in Hyderabad, India.

Nitin Bassi is Senior Researcher with IRAP and is based in Delhi, India.

A. Narayanamoorthy is NABARD Chair Professor and Head in the Department of Economics and Rural Development, Alagappa University, Karaikudi, Tamil Nadu, India.

M.V.K. Sivamohan is Principal Consultant for IRAP, and is based in Hyderabad, India.

Earthscan Studies in Natural Resource Management

1. **The Water, Energy and Food Security Nexus**
   Lessons from India for Development
   Edited by M. Dinesh Kumar, Nitin Bassi, A. Narayanamoorthy and M. V. K. Sivamohan

2. **Gender Research in Natural Resource Management**
   Building Capacities in the Middle East and North Africa
   Edited by Malika Abdelali-Martini and Aden Aw-Hassan

3. **Contested Forms of Governance in Marine Protected Areas**
   A Study of Co-Management and Adaptive Co-Management
   By Natalie Bown, Tim S. Gray and Selina M. Stead

4. **Adaptive Collaborative Approaches in Natural Resource Management**
   Rethinking Participation, Learning and Innovation
   Edited by Hemant R. Ojha, Andy Hall and Rasheed Sulaiman V

5. **Integrated Natural Resource Management in the Highlands of Eastern Africa**
   From Concept to Practice
   Edited by Laura Anne German, Jeremias Mowo, Tilahun Amede and Kenneth Masuki

For more information on books in the Earthscan Studies in Natural Resource Management series, please visit the series page on the Routledge website: http://www.routledge.com/books/series/ECNRM/

The Water, Energy and Food Security Nexus

Lessons from India for development

Edited by M. Dinesh Kumar, Nitin Bassi, A. Narayanamoorthy and M.V.K. Sivamohan

First published 2014 by Routledge

2 Park Square, Milton Park, Abingdon, Oxfordshire OX14 4RN
711 Third Avenue, New York, NY 10017

Routledge is an imprint of the Taylor & Francis Group, an informa business

First issued in paperback 2017

British Library Cataloguing-in-Publication Data
A catalogue record for this book is available from the British Library

Library of Congress Cataloging-in-Publication Data

The water, energy and food security nexus : lessons from India for development / edited by M. Dinesh Kumar, Nitin Bassi, A. Narayanamoorthy and M.V.K. Sivamohan.
pages cm. — (Earthscan studies in natural resource management)
Includes bibliographical references and index.
1. Water-supply—India—Management. 2. Water-supply—Developing countries – Management. 3. Water resources development—India. 4. Water resources development—Developing countries. 5. Food security—India. 6. Food security—Developing countries. 7. Sustainable agriculture—India. 8. Sustainable agriculture—Developing countries. I. Dinesh Kumar, M.

TD303.A1W344 2014
333.70954—dc23 2013029094

ISBN: 978-0-415-73303-8 (hbk)
ISBN: 978-1-138-57476-2 (pbk)

Typeset in Baskerville
by Apex CoVantage, LLC

Contents

For those who are living in ecosystems full of uncertainties in the developing world.

Figures

Tables

Contributors

Nitin Bassi works as Senior Researcher with the Institute for Resource Analysis and Policy (IRAP) and is based at Delhi. He holds an M.Phil. in natural resources management from Indian Institute of Forest Management (IIFM), Bhopal, and is a Ph.D. research scholar with the University of Delhi. He has over 7 years of experience undertaking research in the field of water resource management. His area of expertise include institutional and policy analysis in irrigation and water supply management, agricultural water management and wetland management. Before joining IRAP, he worked with the International Water Management Institute (IWMI) on the issues of farmer participation in irrigation management and groundwater governance in Asia. He has several publications, both national and international, to his credit.

Yusuf Kabir has worked with UNICEF since 2007. Prior to that he worked with organizations like DFID; national-level NGOs; and social and marketing research consultancy firms like GFKMODE, ORG India Pvt Ltd., Ramky Infrastructure, SREI Capital Markets and SPAN Consultancy, on issues related to environmental and livelihood development training and capacity building in the social sector. He holds two postgraduate degrees in sustainable development and environmental management.

M. Dinesh Kumar is Executive Director at the Institute for Resource Analysis and Policy, Hyderabad. Dr. Kumar holds a Ph.D. in water management, an ME in water resource management and a BTech in civil engineering. Dr. Kumar is a senior water resources scientist with over 22 years of experience working on technical, economic, institutional and policy-related issues in water management in India. He has published in several international and national journals on water and energy, and has four books, two edited volumes and over 140 research papers to his credit. Currently, he is also Associate Editor of *Water Policy*, an international peer-reviewed journal published by IWA.

Susanto Kumar Beero is a fellow in the Department of Economics and Rural Development, Alagappa University, Karaikudi, Tamil Nadu.

A. Narayanamoorthy is currently working as NABARD Chair Professor and Head, in the Department of Economics and Rural Development, Alagappa

University, Tamil Nadu, India. Previously, he worked at the Gokhale Institute of Politics and Economics, Pune, for about 13 years (1994–2008) and at the Madras Institute of Development Studies, Madras, from 1993 to 1994. His area of work is in agricultural and rural economics. His specialization is in irrigation. Dr. Narayanamoorthy has published five books, and over 90 research papers in international and national journals. He has piloted several research projects sponsored by the Ministry of Agriculture (Government of India), Planning Commission (Government of India) and the National Bank for Agriculture and Rural Development (NABARD), India.

V. Niranjan works as a research officer at the Institute for Resource Analysis and Policy, Hyderabad. He holds an M.Tech in environment management from Jawaharlal Nehru Technological University, Hyderabad.

Neena Rao is Director – Projects and Partnerships and South Asia Regional CapNet UNDP Network (SCaN) Manager, SaciWATERs. Currently, she is also a board member of the CapNet, UNDP Global Network for Training and Capacity Building in IWRM. Dr. Rao has varied and diverse national and international experience in research, training and implementation in the development field and has wide experience in working with diverse ethnic groups in the United States and India. Her interests are in systems research, training, design and implementation of various projects in the development field. She has several national and international publications to her credit.

K. Siva Rama Kishan has a masters degree in anthropology from University of Hyderabad, and is currently working as a research officer at IRAP.

Christopher A. Scott is Associate Research Professor of water resources policy at the Udall Center for Studies in Public Policy, with a joint appointment as Associate Professor in the School of Geography and Development, University of Arizona, United States. His work focuses on the policy dimensions of global change and water (climate change and urban growth). Dr. Scott holds Ph.D. and M.S. degrees in hydrology from Cornell University. He has over a decade of rich experience living and working in India, Nepal, Sri Lanka and Mexico (with the International Water Management Institute), and working with non-governmental organizations.

O. P. Singh has obtained his M.Sc. (math) from Banaras Hindu University, his A.M. from the University of Pennsylvania (USA) and his Ph.D. from Banaras Hindu University. His fields of specialization are distribution theory, pseudo-differential operators, wavelets and nonlinear analysis. He is a member of several professional societies.

M. V. K. Sivamohan is Principal Consultant to the Institute for Resource Analysis and Policy, Hyderabad. He has over 40 years of research, consultancy and training experience. Formerly a senior member of the faculty and area chairman (agriculture and rural development) at the Administrative Staff College of India, Hyderabad, Dr. Sivamohan holds a Ph.D. in the social sciences. He has worked with several international organizations, such as the Irrigation Research Group, Cornell University (USA); the Natural Resource Institute (UK); and the International Water Management Institute (India). He has several publications in both national and international journals and four edited volumes to his credit. His specialization is in natural resources management and public administration.

L. Venkatachalam is Associate Professor at the Madras University of Development Studies (MIDS), Chennai. His areas of interest include environmental economics and behavioral economics. He was a Fulbright senior researcher at the department of Agricultural and Resource Economics, University of California, Berkeley, United States. He has published profusely in the area of agriculture economics.

Foreword

No rising issue demands attention more than the growing concern for food security, especially in poverty-stricken developing countries. The editors of this text have put together a carefully coordinated team of authors who have empirically studied key aspects of the nexus of water and energy that threaten the sustainability of the food system. The authors' experiences represent local and regional studies from around the globe. When lessons from these studies are integrated, they point toward new and better approaches for co-managing water and energy to achieve more secure food systems.

In each chapter, authors share their experiences with specific case studies looking at how resources, including water, energy and land use for agriculture, are influenced by policymakers, politicians, agro-businesses, corporations and others. Some chapters focus on lessons learned from specific, often notable projects or programs. Examples of these include the analysis of the 'miracle growth' in agricultural production in Gujarat; the implementation, success and failure of large government employment schemes; hydropower generation as an engine for economic growth; relevancy of irrigation development on rural poor; and the impact of the emerging big agro-businesses on water and food security. With sustainable water supply a key determinant of sustainable agriculture, the authors look comprehensively at use of various inputs such as energy sources (e.g. diesel or electric), water sources (e.g. groundwater or surface water, local or regional), management approaches (e.g. private or government), agricultural technologies (e.g. trickle or canal irrigation), etc., any and all of which can influence the outcome for food security.

The book draws from empirical data of specific projects including some groundbreaking case studies done in south Asia, all related in important ways to the nexus of water and energy and the impact of that nexus on agriculture and the resulting security of the food system. Competition for water and energy ultimately affects not only the food system but also the sustainability of the environment, ecosystems and the water resource itself. The very idea of sustainability is to provide for future generations. This book clearly shows that we are on a trajectory where the very building blocks for human and environmental preservation are being used inefficiently and wastefully. Fortunately, the book also provides a framework of how to effectively begin to turn this trajectory into a more sustainable future for all life.

The editors of this book are all world-renowned scholars from India. Three of the editors are associated with the Institute for Resource Analysis and Policy in Hyderabad: Dr. M. Dinesh Kumar, who is the executive director of the institute; Mr. Nitin Bassi, a senior researcher; and Dr. M. V. K. Sivamohan, a principal consultant. From Alagappa University in Karaikudi, Dr. A. Narayanamoorthy, professor and director, is the fourth editor. These four scholars have a rich record of scholarship related to resource analysis and policy, especially as it relates to the water-energy nexus and food security. These editors also joined other distinguished scholars in coauthoring chapters of the book.

The book starts with an excellent overview of the increasing demands of water, energy and other resources; the reason for these new pressures; and why this development is putting the international community goals of economic development, improvement in human well-being, and poverty eradication in jeopardy. The first chapter introduces discussion of several rising paradigms for developing countries based on their climate, economy and management limitation to achieve water, energy and food security. The final chapter pulls together many of the lessons learned from the more focused studies of earlier chapters and draws from them useful ideas for developing counties to comprehensively manage their water and energy for better food security. The authors also introduce a new nexus of 'politics-bureaucracy-academics,' which is leading to 'huge negative impacts on the management of natural resources and their economies.'

Because this book provides such an interdisciplinary view of how a sustainable food system depends on responsible policies and management of water, energy and other inputs, it will be useful to people from many walks of life, including students, academics, policymakers, consultants and industry. It is a book that can be read cover to cover or used as a reference for many critical food security topics related to water, energy and a host of issues that link the two.

Prof. Michael Faiver Walter
Cornell University

Acronyms and abbreviations

ANA National average in rural poverty
APHW Asia Pacific Association of Hydrology and Water Resources Conference
ARWSP Accelerated Rural Water Supply Programme
BNA Below national average in rural poverty
BTU British thermal units
CAFOs Concentrated animal feeding operations
CT Cash transfer
EGS Employment guarantee scheme
EMC Multidisciplinary environmental monitoring committee
ET Evapo-transpiration
FPPHRP Food grains production (kg) per head of rural population
GCA Gross cropped area
GIA Gross irrigated area
GIS Geographical information system
GOI Government of India
GOM Government of Maharashtra
GP Gram panchayat
GS Grama Sabha
GSDA Ground Water Survey and Development Agency
GW Giga watts
GWB Central Ground Water Board
HDI Human development
IAPTRP Irrigated area (ha) per thousand rural population
ICAR Indian Council of Agricultural Research
IEA International Energy Agency
IGP Indo-Gangetic Plain
IHA International Hydropower Association
IRR Initial irrigation rate
kWh Kilowatt hours
MGNREGS Mahatma Gandhi National Rural Employment Guarantee Scheme
MJP Maharashtra Jeevan Pradhikaran
MWSSB Maharashtra Water Supply and Sewerage Board

MWSSB Act	Maharashtra Water Supply and Sewerage Board Act
NGO	Nongovernmental organization
NRDWP	National Rural Drinking Water Programme
NREGA	National Rural Employment Guarantee Act
NREGS	National Rural Employment Guarantee Scheme
NRLP	National River Linking Project
O&M	Operation and maintenance
PES	Payment for ecosystem services
PFG	Productivity (kg) of food grains
PHE	Public health engineering
PRP	Percentage of rural poverty
RSMU	Reform Support and Project Management Unit
RWAL	Real wage rate of agricultural laborers
RWSS	Rural Water Supply and Sanitation
SADC	Southern African Development Community
SDAPHRP	State domestic product (Rs) of agriculture per head of rural population
SDP	State domestic product
SEB	State electricity boards
SOFILWM	Society for Integrated Land and Water Management
SRP	Sector Reforms Project
SSP	Sardar Sarovar Project
SWI	Sustainable water use index
SWID	State Water Investigation Department
TFP	Total factor productivity
UP	Uttar Pradesh
VWSCs	Village water and sanitation committees
WB	West Bengal
WBPDCL	West Bengal Power Development Corporation Ltd.
WBSEDCL	West Bengal State Electricity Distribution Company Ltd.
WCD	World Commission on Dams
WENEXA	Water and Energy Nexus Project
WHS	Water-harvesting structures
WM	Water management
WSSD	Water Supply and Sanitation Department
WSSSO	Water and Sanitation Support Organization
WWRC	Watershed Water Resources Committee
ZP	Zilla Parishad

Preface

This book is about managing water and energy for sustainable water, food and energy security in developing countries, drawing upon the evidence based on research in India. For developing economies, the ability to maintain good economic growth rates and move people from poverty depends heavily on how water and energy resources are managed for sustainable water supplies, and food and energy security. This requires solutions to several complex problems relating to the latter. Along with good technology and human capital, this would more often require hard policy decisions on the part of the state bureaucracies dealing with water, energy and agriculture, supported by strong political leadership. Such policy decisions have to be supported by sound understanding of the working of water-energy-food systems, generated through evidence-based research.

A lot of misconceptions exist about how water and energy resources can be managed for addressing long-term water, energy and food security, primarily due to lack of adequate research in the field. In the context of water supply, a few of these misconceptions are that the days of large irrigation infrastructure are over, as they no longer generate attractive returns against investments and that small water-harvesting structures would help improve water security for the poor in water-scarce regions, through supplementary irrigation, drinking water supply augmentation, etc. Another misconception is that countries like India should improve water and food self-sufficiency through virtual water imports, rather than using internal (water) resources for producing food domestically at the cost of ecology and the environment. While the government of India continues to invest gargantuan sums on water harvesting every year through social welfare schemes, there is hardly any scientific evidence to show the positive impacts of these schemes on water availability at the local and basin level, and on rural livelihoods.

The energy sector is also plagued by several such misconceptions, leading to detrimental policies and undesirable consequences for water, energy and food security. For instance, it is well known that the demand for energy is growing in developing countries like India, with a widening gap between supply and demand. It is also well known that energy use efficiencies remain very low in agriculture, which is one of the largest users of energy in these countries, primarily due to heavy subsidies on electricity and diesel. But the potential for energy-demand management through efficient pricing has hardly been explored because of considerations of political economy. Both energy and water are wasted, especially in the agriculture sector, producing poor returns. With declining revenues, such policies affect the viability of the power sector adversely, leaving the utilities with little resource for investment, including that for power generation and distribution. Wide gaps exist between the country's hydropower potential and the present utilization. As a consequence, the quality of power supplied to agriculture and other sectors of the economy suffered severely during the past few years in many Indian states. Diesel subsidies make a huge dent on government finances.

The widely held view is that raising energy tariff, be it for electricity or for fossil fuel, would hurt the livelihoods of the poor most, especially because they heavily depend on these forms of energy for irrigated agriculture. The real impact of energy pricing on economics of farming, and energy and water use efficiencies, had not been systematically studied based on empirical work. Plans for investment in cleaner power projects are adversely shrouded in controversies due to growing opposition from environmental lobby for ecological damage and human displacement. This is even more so in the case of hydropower. The view that hydropower dams in regions like the northeast, which has enormous generation potential, can only spell doom for ecology and human safety, is part of 'environmental fundamentalism,' and not based on facts about the performance of hydropower dams in the world. Nevertheless, we need to strike a balance between local peoples' needs, concerns and priorities including those that are ecological, and the larger developmental goals. While several of the papers presented in this book challenge the long-held views about water-energy-food security nexus, they also offer some concrete solutions to the problems of growing water and energy scarcity to ensure sustainable water supplies, energy security and food production.

We thank Professor Michael F. Walter, Jr., for writing the foreword to this book. We are thankful to Tim along with Ashley, Deepti and other team members of Earthscan/Routledge for their support in publishing this work and prompt advice when needed. Thanks are due to the contributors to the book chapters and also the anonymous reviewers of this book.

M. Dinesh Kumar
Nitin Bassi
A. Narayanamoorthy
M. V. K Sivamohan

1 Water-food-energy nexus

Global and local perspectives

M. Dinesh Kumar, Nitin Bassi, A. Narayanamoorthy and M. V. K. Sivamohan

Introduction

Access to clean drinking water, energy and food at a level sufficient to meet basic human needs is a fundamental right or, in a way, directly promotes those rights. Global demands for water and energy resources and food are set to increase dramatically, as a result of rising incomes, demographic shifts and changing consumption patterns and lifestyles. The basic resources for producing energy and food and provision of water supplies are under increasing stress. This is endangering water, energy and food security, putting in jeopardy the development goals of improved human well-being, economic development and poverty eradication.

It is estimated that there will be a 40% overall gap between global water supply and demand by 2030. If current trends continue, by 2030 two-thirds of the world's population will live in areas of high water stress (The 2030 Water Resources Group, 2009). According to estimates by International Energy Agency, by 2030, overall demand for energy will increase by around 40% from the 2010 levels (World Economic Forum, 2011), rising from around 500 quadrillion British thermal units (BTU) to around 700 quadrillion BTUs, even with significant improvements expected in energy use intensity across sectors (ExxonMobil, 2013). Further, the demand for food could be 70% higher than the 2005/07 levels by 2050 (FAO, 2009).[1] Changing diets – driven by rising incomes and other shifts – will increase the demand for resource intensive farm products such as meat (World Economic Forum, 2011). Both rural and urban areas will face challenges for access to basic services relating to water and energy.

Unfortunately, the distribution of these resources will be uneven across different regions and countries – developed to developing, tropical to temperate, arid to humid. The developed countries have already obtained high levels of water (Grey and Sadoff, 2007; Kumar, 2009) and food security (Kumar, 2009; IFPRI, 2011). As regards energy security, high-income countries have achieved energy security as well, with much higher levels of electricity consumption in terms of kilograms of oil equivalent per capita (4,856) as compared to the world average (1,788), which is far higher than the developing countries like India and China (World Bank, 2012).[2] Major improvements in water supply, energy security and food supplies are yet to be achieved in the developing world. This would be a key determinant of future growth in demand alongside population growth, and therefore scarcity.

The need to increase food production to meet the growing demands would considerably increase the demand for irrigated water and energy in tropical climates, which require extensive artificial application of water for crop production. In the semi-arid and arid tropics, the energy demand for irrigation would be even higher, as compared to humid and sub-humid tropics, as the dependence on groundwater for irrigation is more acute in the former. Unfortunately, major shares of the world's uncultivated arable land are in the semi-arid and arid tropics of Africa (Svendsen et al., 2009) and Asia (FAO, 2000), and the crop lands that require irrigation are also mostly in Africa (Svendsen et al., 2009) and Asia. Since a large number of the world's poor and the malnourished people live in these regions, this essentially means that a significant proportion of the future increase in water and energy demand for food production is likely to be from these regions, and this demand growth per every unit increase in population would be far higher than that of the past due to climatic factors. Added to this problem, population growth rate, a major driver of their demand, is unlikely to be uniform, and is higher in developing countries.

The manner in which the increased demand for food would drive the demand for water would also be determined by the efficiency of water use in the future for agricultural production, the major consumptive user of water globally. Similarly, the way increased crop production would increase the energy demand would again be influenced by the technical efficiencies in energy use in that sector in the coming years. In regions where the linkage between energy and irrigation is strong,[3] the way water is managed and used in agriculture would also determine the ability to provide reliable and adequate supplies of energy for other sectors of the economy (Kumar et al., 2013b). Here again, the overall approach for sustainable groundwater management and approaches for water demand management in agriculture, which include technological, institutional and policy options, would be the key. Reforms in the energy sector, which encompasses technologies, institutions and policies, would have a big role in managing the energy demand. In any case, increased demand for water in agriculture, which is not matched by significant efficiency improvements, would result in direct competition between water use in agriculture and domestic water supplies in regions where groundwater is the major source.

Increased global, national and regional demand for energy can exert pressure on precious water resources due to their diversion for production of biofuel (Earth Policy Institute, 2011) and use in thermal power plants. Studies show that a typical fossil fuel-based thermal power plant can consume around 3.5 cubic meters of water per megawatt of power generated (The United States Department of Energy, 2005), while the total water required to be diverted would be much larger.[4] This can put scarce water and land resources under enormous stress, leading to undesirable effects on food production systems, unless both land and water are in abundance. While in naturally water-scarce areas with abundant arable land, the pressure would be on water resources, in water-rich areas with limited arable land, the pressure would be on land. Therefore, the impacts on food production systems would differ from region to region depending on the climate and availability of water resources and arable land.

Different aspects are intertwined in the 'food-energy-water' nexus. One of the most important aspects, which we will deal with in this book, is sustainable use of water and energy resources to meet the growing food security and agricultural growth challenges in the developing world. We will illustrate through case studies how efficient management of energy economy through efficient pricing and rational supply policies for energy will promote efficient and sustainable water management. We will also show why sustainable management of water resources is crucial for managing the energy economy in regions where marginal returns from the use of water and energy is very high in agriculture. Further, we will show how global concepts have been misused to create illusions about the heightened role of irrigation and energy security in promoting agricultural development and food security in certain water-rich regions of the world in order to support pervasive policies of subsidizing electricity.

The rationale for the book

Over the past 2 decades, the water management debate in many developing countries in the tropical climate, which are facing acute shortages of water for food production and basic survival needs, had been dominated by two new paradigms. First is the decentralized approach to water management, involving construction of water harvesting and groundwater recharge structures (Kumar, 2010). The underlying premise is that while rain occurs everywhere in the basin, the systems to capture it can also be decentralized to make it socially viable and managerially efficient, and that it would involve no major land submergence and displacement of people living in the catchments, and no social and ecological consequences. The growing problem of depletion of groundwater resources in many arid and semi-arid regions (World Bank, 2010) seems to have provided a strong economic and ecological rationale for such interventions (Shah et al., 2003). Ecosystem health was defined in a narrow perspective of leaving large rivers untouched by human interventions. The fact that many small flowing streams contribute to the flows in large river systems has been conveniently ignored due to lack of basin perspective among the planners of decentralized water systems.

The second paradigm is the 'soft path' of virtual water import to achieving total water sufficiency, in order to save precious water resources from being diverted for agriculture in water-scarce regions (Kumar, 2010). Some international development scholars used the concept of virtual water trade, a term coined by Professor Tony Allan in the early 1990s, to fiercely argue that naturally water-scarce regions should import food, with embedded water, from water-rich countries in order to save their own precious water resources, by highlighting the virtues such as 'global water use efficiency,' 'distribution of scarcity' (Allan, 1997; Allan, 2003), and averting 'water wars' (Allan, 2003). But, there is ample evidence available from empirical research that the virtues of global water use efficiency and distribution of scarcity are not achievable at the operational level through virtual water trade, and that globally virtual water flows from water-scarce and land-rich regions to water-rich and land-scarce regions.

Being subject to growing pressure from international donors, who express concern over the inability of national governments of the developing world to address the potentially devastating social and environmental consequences of large water resource systems, the development practitioners and policymakers alike have found it politically convenient to pursue decentralized water management solutions in the water sector.

The government of India has been investing nearly $8 billion annually, for the past 8 years, for a social protection program called National Rural Employment Guarantee Scheme (NREGS). But the design of the scheme is conceptually flawed. It fails to take into account the fact that the rural employment scenario varies vastly across India, with some regions having excessively high rural unemployment and poverty rates, while some other regions have very high demand for rural wage laborers, and that the employment generation potential of regions with high unemployment rates through public works is extremely poor. Further, in spite of mounting evidence to the effect that the water harvesting systems built in the naturally water-scarce regions of India perform poorly in both hydrologic and economic terms (Kumar et al., 2006; Kumar et al., 2008), the planning of water harvesting and conservation structures under NREGS does not take into account the hydrology and topography of the regions concerned, which have serious implications for physical benefits and cost effectiveness.

The major critiques of the scheme have focused on the problems in implementation (Ambasta et al., 2008), with overwhelming emphasis on the issues of fund embezzlement and corruption and the ways to tackle them (Adhikari and Bhatia, 2010; Vanaik and Siddhartha, 2008). Very little attention has been paid to the conceptual aspects of the scheme design, though a recent evaluation of the scheme critiqued it for selecting certain water-related works without giving due consideration to the agro ecology and topography of the localities concerned (see, for instance, NCAER/PIF, 2009). The limited empirical studies, which had eulogized the scheme, looked at villages as their unit of analysis for analyzing the socio-economic and environmental impacts (see Tiwari et al., 2011). Notwithstanding the methodological flaws in carrying out such impact assessment, by no standard can a village become the unit for analyzing the impacts of water interventions, when they are carried out at a massive scale.

Parallel to the virtual water trade argument, there has been a growing orchestration in the development circle that the water-rich regions of the developing world could boost their agriculture production if pumping equipment and sufficient amounts of energy are made available to the farmers for abstracting cheap groundwater available at shallow depths. Interestingly, many water-rich regions are inhabited by poor communities. The eastern Indo-Gangetic Plains (IGP), which constitute the 'poverty square of South Asia' is one of the most water-rich regions in South Asia. Yet, it is one of the most food-insecure regions in the entire world (Kumar et al., 2013a; Sikka and Bhatnagar, 2006). It is also characterized by the lowest level of water in agriculture. Hence, the argument was extended to the eastern Indo-Gangetic Plains as well (see Sikka and Bhatnagar, 2006). However, such debates have never considered the fact that climatic (Aggarwal et al., 2000;

Pathak et al., 2003) and ecological factors (Ladha et al., 2000) become constraints to raising yields of rice and wheat, the dominant crops of eastern IGP, to levels comparable with that of the highly productive regions of Punjab.

Another important factor, which is critical to the agricultural and regional economic growth of eastern India, is energy security. The region has a huge electricity demand in the agriculture sector, and in the absence of adequate power generation and distribution infrastructure, this demand is partly met by millions of diesel-powered engines. Such an energy crisis in eastern India occurs in spite of the huge untapped (8%) hydropower potential in the neighboring northeastern region.

Subsidized electricity supply to the farm sector, provision of free power connections to the small and marginal farmers, who are dependent on water purchase or 'costly' diesel-pump irrigation (Mukherji, 2008; Mukherji et al., 2012) and subsidized diesel pump sets and fuel for agricultural use (Shah, 2001) are the standard prescriptions of some academics and researchers for boosting agricultural production in this region and for moving people out of poverty. Such arguments carry high political currency. The recent decision of the West Bengal government to launch a scheme that offers free power connections to farmers in the state had been touted by many as an effective public policy intervention to boost irrigated agriculture and reduce rural poverty. However, this decision was not based on any analysis of the economic viability of the scheme at the societal level and its equity implications – particularly who would ultimately benefit from such a scheme, whether the water buyers or the rich diesel well owners.

The intensity of use of fossil fuel or electrical energy for crop production per farmer and per unit of crop land is perhaps one of the lowest in this region, owing to the shallow groundwater table, the low per capita arable land and the relatively higher rainfall and subtropical climatic conditions. Yet, at the aggregate level, it constitutes one of the largest users of fossil fuel for crop production, given the high population density in rural areas that are agrarian. It is estimated the 6 million agricultural pump sets[5] run by diesel engines in that the region use around 2,400 million liters of diesel every year (see Chapter 7 of this book). A lot of this fossil fuel is burned in irrigating water-intensive wheat and paddy, through highly inefficient pump sets. Farmers continue with such low-valued crops because of the stable procurement prices offered by the government, in the absence of infrastructure for marketing high-valued crops. The subsidized prices farmers pay for diesel is another reason. The problem of poor pump efficiencies is compounded by poor energy use efficiency in physical terms and water use efficiency in physical and economic terms in crop production. Lower crop outputs in value terms and higher energy footprint in agriculture are the undesirable outcomes.

It is imperative that for the developing countries of the tropics, where a major portion of the population is heavily dependent on agriculture as a primary source of occupation, energy security, which is essential to propel economic growth, is directly linked to how they manage their fossil fuel economy. Efficient pricing of diesel is an integral part of managing fossil fuel economy. But, the dominant argument in favor of diesel subsidies is that removing diesel subsidies would force farmers out of agriculture, as the cost of irrigation would go up, with impact not only on the diesel well owning farmers, but also the millions of farmers who purchase water from diesel pump owners or hire diesel engines. This is politically convenient, as moving out the subsidy regime means that farmers are provided with the infrastructure to market the high-valued crops, which they would produce in the event of rising energy costs for irrigation.

The water-energy nexus is different in semi-arid water-scarce regions. In these regions, agricultural production is water intensive (Kumar et al., 2012). Also, groundwater use for agricultural production is energy intensive, owing to deep water table conditions as a result of overexploitation of the resource. Electricity supply is not only heavily subsidized in most states falling in these regions, but also is unmetered. In cases where the electricity supply is not free, the electricity utilities charge for agricultural power consumption on the basis of the connected load. The higher prices for efficient pump sets and flat rate charges for electricity encourage farmers to use inefficient pump sets (Singh, 2009).[6]

The political economy of power pricing characterized by the political compulsions to continue with subsidies (Singh, 2009; Shah et al., 2004) and poor willingness on the part of farmers to pay the cost of electricity (Shah et al., 2007), and the high transaction cost of metering associated with millions of rural electric power connections for farms and the difficulty in accessing remote villages (Shah et al., 2004; Kemper, 2007) were cited as reasons. The other reasons cited against pro rata pricing are reduced economic returns from irrigated crop production (Saleth, 1997) and reduced welfare gains of high energy prices (Shah, 1993), though the validity of these arguments were not tested empirically. The government of India spends a whopping $6–7 billion annually toward electricity subsidy in the farm sector. The result is that the marginal cost of using electricity and water is almost nil for the well irrigators, leading to overexploitation of aquifers and wasteful use of water in crop production. The long-term impacts of such pervasive policies, being pursued by successive governments, on sustainability of agricultural production and viability of power sector are serious.

While water for agriculture is crucial for food security, economic growth and reduction of rural poverty in the developing world, household level water security is equally important for bringing about social advancements including improved health and well-being, education and building a productive workforce (HDR, 2006), though in volumetric terms water for household uses would account for a fraction of the water for irrigation. In recourse to improving water security at the household level, the governments in developing countries around the world are under enormous pressure to make massive budgetary allocations for provision of rural water supply services, in the form of infrastructure costs and operation and maintenance of schemes. In spite of the perceptible social benefits from improved domestic water supply, the value communities attach to these services is often low, reflected in the ridiculously low price that they are willing to pay for it, against the actual cost of the service incurred by the public agencies. This is against the backdrop of communities in similar socio-economic settings elsewhere incurring substantial costs for purchase of good quality water from private vendors. This forces the government to heavily subsidize water from such schemes. The poor financial viability of rural water supply schemes in turn seriously inhibits the ability of these utilities to maintain the basic level of services – with adequate staff and resources for continuous upgrading of the infrastructure to meet the growing demand or the maintenance and upkeep of the existing ones.

The reason is that the rural water supply schemes are generally planned as 'single use systems,' for meeting only the domestic water supply needs of the population. But a rural population has many productive water needs.[7] Water supply systems that do not consider the needs of rural communities for sustainable livelihoods fail to play an important role in their day-to-day life. During scarcity, the poor communities often compromise their personal hygiene requirements in order to meet water requirements for productive needs. Failure on the part of the water supply agencies to maintain supply levels that cover productive as well as domestic needs would result in the households not being able to realize the full social and health benefits from water.[8]

Another possible outcome is that the system by default becomes a multiple use system. While some of the unplanned uses may get absorbed by the system, other uses can damage it. This is compounded by the often intermittent and unreliable nature of water supplies. As the rural communities are not able to perform economic activities from the water supplied through public systems, they show low levels of willingness to pay for the water supply services. This affects the sustainability of the systems, as official agencies are not able to recover the costs of their operation and maintenance.

But, whenever such unplanned uses take place from 'single use systems' without causing much damage to the physical infrastructure, it brings about improvements in water-related livelihoods through freedom from drudgery, health improvement, food production and income. This leads to the point that a marginal improvement in drinking water supply infrastructure and a marginal increase in the volume of water supplied can enhance the value of water supplied remarkably in social as well as economic terms.

However, planning of water supply systems for multiple uses is restricted by lack of comprehensive data on a wide range of variables, which are vital for designing such systems. They include the multiple water use needs of the households in a locality; the households that are actually vulnerable to problems associated with lack of water for multiple needs; the methodologies for estimating household water demand for meeting domestic and productive needs; and the type of technological interventions feasible in a region for augmenting the resource in terms of quantity and quality for meeting the additional water needs, as well as retrofit systems for improving the access, reliability and equity in water supply.

The foregoing analysis suggests that the developing countries in the tropics still need to invest in water storage and power infrastructure, though their water and energy footprints per unit of agriculture production are very high. But the systems for delivering water to rural areas need to be designed as multiple use systems in such a way that they take care of the range of domestic and productive water needs of the local communities if they have to be sustainable in the long run. These countries need to design policy instruments for effectively managing the demand for water and energy in agriculture to become water, energy and food secure.

The purpose of the book

The book is about sustainable management of water and energy resources in the developing world for agricultural growth, water, energy and food security and rural livelihoods. Using global data sets, it shows that improving water security in general would drive economic growth of nations through advancements in human development and reduce human poverty and income inequity. In arid and semi-arid tropical countries, enhancing water storage in per capita terms is crucial to improving water security.

But it also shows that building too many small water-harvesting systems across a basin is not an oxymoron for enhancing per capita storage. Our analysis instead shows that decentralized water-harvesting activities such as capacity enhancement of local water bodies and construction of small water-impounding structures built under the rural employment guarantee schemes, without due consideration to the regional/local hydrology, topography and agro-ecology, in naturally water-scarce regions do not produce any positive welfare impacts, and instead would distort the local labor markets, and reduce inflows into reservoirs located downstream of these local catchments, with overall adverse effect on the hydrological balance.

The empirical analysis presented in the book shows the enormous scope for co-management of water and energy in developing economies where both are scarce at the national level for fostering water and energy security, and agricultural growth through appropriate technological interventions and market instruments.

In regions where energy is available for supporting irrigated crop production but water is scarce, metering and pro rata pricing of electricity in the farm sector and maintaining its quality and reliability would help improve not only the energy economy, but also agricultural economy based on groundwater use, as famers improve physical efficiency of water use and modify their farming systems to include water-efficient crops, thereby enhancing their farm returns and reducing groundwater use. It substantially reduces the energy footprint involved in agriculture.

In contrast, in regions that have abundant water resources at the societal level, but where the government and the local communities lack the wherewithal to exploit it in an efficient manner, the standard prescription has been to offer free power connection and supply. But the analysis presented in the book questions this approach by taking the case of the recent public policy initiatives of the eastern Indian state of West Bengal. Empirical analysis involving data on the agrarian structure, access to irrigation and power supply infrastructure and agro climate shows how these lopsided policies, which do not take cognizance of the agrarian structure of the region, allow the rich farmers to continue exploiting the resource-poor small and marginal farmers, who already experience inequity in access to groundwater resources, using their 'monopoly power' by cornering the subsidy benefits.

The book also discusses how it is important to extend the concept of household level water security in the rural context to cover productive as well as domestic water needs to help improve livelihood and nutritional security. But not all households have the same degree of vulnerability to inadequate water supply for multiple needs. How far a household is vulnerable is determined by the socio-economic profile of the households, the physical environment – hydrology, climate and drought proneness – and cultural settings and institutional regime. Therefore, a composite index that helps assess the vulnerability of individual rural households is discussed. The book also covers the design of multiple use water systems for reducing this vulnerability for three regions in Maharashtra, which are characterized by differences in physical, socio-economic and cultural settings.

In a nutshell, the book provides policy analysis and helps in future policy formulation by national governments and local institutions. The book deals with wide-ranging topics on these issues, trying to analyze and link them coherently for better informing of some crucial aspects regarding the topic of the water-food-energy nexus.

Contents of the book

Gujarat, in western India, historically known for poor endowment of water and land resources and with severe droughts experienced frequently in most parts, strikingly demonstrated the impact of large water and energy systems on economic growth. The state had been eulogized by many researchers, policymakers and politicians for the bold initiatives taken by its government to boost agriculture, departing from this track, which according to them led to a miracle growth of 9.6% per annum (Shah et al., 2009). Gujarat is known for intensive use of groundwater and energy for crop production, and is historically known for intensive dairy farming. According to some researchers, the 'Gujarat model for agricultural growth' is characterized by some distinct features: 1) high-quality, uninterrupted power supply to the farm sector; 2) intensive, but decentralized, small-scale water harvesting, with community participation; 3) a special purpose vehicle created by the state government, called 'Gujarat Green Revolution Company,' to promote micro irrigation systems in the state, as a water- and energy-saving system for the farmers; and 4) an innovative agricultural extension program launched by the state, called '*Krishi Mela.*'

Chapter 2 critically examines the claim about the miracle growth, by analyzing the data on agriculture in the state over the past 60 years from 1951–2010, with detailed analysis of major crop segments (wheat, groundnut, cotton and paddy) and the dairy-farming (milk) sector. Simultaneously, it also examines whether the much-hyped initiatives of the state government have the potential to bring about changes in the state's agriculture, with the help of studies in the recent past on these initiatives. Through the analysis, it brings out the distinct features of agricultural growth in the state, identifying the time periods during which the state's agriculture sector recorded real and consistent growth and identifying the factors that really determine the growth. The analyses provide some important lessons for water management in other semi-arid and arid regions of the country, not only because of its implications for agricultural production, but also due to its positive linkage with advancements in human development and economic growth.

The water management component of NREGA has drawn on the positive experience of many states like Gujarat and Rajasthan in small water harvesting. It is being lauded by many in the academic, development and policy arena as a 'silver bullet' for eradicating rural poverty and unemployment, by way of generating demand for productive labor force in villages and private incentives for management of common property resources. Chapter 3 reviews the water management activities being executed under the scheme vis-à-vis their effectiveness in improving water security in rural areas of the country, by taking cognizance of the hydrological regime and socio-economic realities of different regions, particularly those that are naturally water scarce. The chapter also reviews the design of the scheme vis-à-vis job entitlement, in terms of its potential to generate jobs and impact the labor dynamic in different situations, by taking into account the vastly varying employment situation across India.

India's farm sector sustains livelihoods for hundreds of millions of rural people, but faces serious management challenges for land, water and energy resources. Growing dependence on groundwater threatens water resource sustainability and power sector viability. Sustaining India's rising prosperity rests on managing groundwater. Many researchers from the region believe that metering electricity use in the farm sector and charging for power on the basis of consumption, which is crucial for managing groundwater and energy economies efficiently, will be political suicide for the ruling governments, as it would be met with resistance from farmers due to reducing profit margins.

Like water, India's per capita energy consumption has grown dramatically – from 1,204 units in 1970–71 to 4,816 units in 2010–11. But the actual demand for power is more than the supply, with the energy shortage being around 7.3%, and shortage in peaking capacity being 11% (Rao, 2006). The conventional fossil fuel–based power generation sources face stiff opposition from environmental lobbyists. Hydropower is the cleanest form of energy, and can also handle peak power demands. But only about 19% of the total hydropower potential of 84,044MW in India (assessed at 60% load factor) had been utilized up to 2005.The northeast has nearly 38% of India's hydropower potential. But as of 2005, more than 98% of this potential remained untapped.

Chapter 4 tries to address whether hydropower generation in the northeast can be an engine for economic growth and development. It states that the hydropower generation in the northeast is being viewed as a panacea for the energy security problems of the country in general and a valued engine for economic growth of states like Sikkim in particular. Consequently, there are a burgeoning number of hydropower projects being planned in the northeast. The government of Sikkim has plans for building 30 such projects on the River Teesta alone. In this context, the chapter addresses the following questions: To what extent is it going to be the growth booster of the region; who are the real beneficiaries, and what does it really mean to the local communities in terms of their livelihoods and risks; and will these projects be ecologically sustainable and economically viable? Further, the chapter presents an analysis of the situation with an interdisciplinary approach toward addressing these questions. It will identify the gaps in the current hydropower development program in the northeast that are currently preventing the program from being an ecologically sustainable, socially just and economically viable solution to meeting India's growing energy demand.

Chapter 5 revisits the debate on irrigation versus poverty by examining the macro level data on irrigation and rural poverty rates in India. The chapter highlights that besides increasing the cropping intensity, adoption of new technology and productivity of crops, the intensive cultivation of crops due to timely access to irrigation increases the demand for agricultural laborers and hence wage rates for those who are mostly living below the poverty line. Both increased affordability of food grains and wage rates help the rural poor to cross poverty barriers. The chapter points out that although the agricultural growth has been used as one of the principal explanatory variables in most poverty-related studies in India, the importance of irrigation has not been recognized by most studies focusing on rural poverty. Thus, the chapter presents analysis on how the availability of irrigation has affected the rural poverty in India using statewide cross-section data covering eight time points. The chapter offers some policy recommendations for future investments in irrigation.

Chapter 6 presents the empirical results from a study carried out in three regions in India – eastern Uttar Pradesh, south Bihar and north Gujarat – on the impact of power pricing on sustainability of groundwater use in agriculture and economic prospects of farming, along with efficiency in use of water for crop production. It shows that raising power tariffs in the farm sector to achieve efficiency and sustainability of groundwater use is both socially and economically viable. The question is about how to introduce this shift. The chapter discusses five different options for power supply, metering and energy pricing in the farm sector and the expected outcomes of implementing each vis-à-vis efficiency of groundwater and energy use, equity in access and sustainability of groundwater.

Though the poverty reduction capacity of agricultural growth in India (Datt and Ravallion, 1998) is being questioned in the light of the declining contribution of agriculture to national economies of many south Asian countries (see Byerlee et al., 2005), this argument in development approach that agricultural growth can reduce poverty is still applicable to many backward regions that are still agrarian (Kumar et al., 2010). In water-abundant eastern India, which has the lowest agricultural growth in the entire country (Evenson et al., 1999) and highest rate of rural poverty, the resource-poor, small and marginal farmers pay exorbitant prices for the water they buy from well owners, while several depend on costly diesel pump irrigation, making irrigated agriculture an unattractive proposition (Kumar, 2007).

Over a period of 16 years, from 1990 to 2006, the price of diesel has gone up from Rs. 5 per liter to Rs. 34.84 per liter (Kumar et al., 2010). Researchers have argued that this has badly hit agricultural growth in eastern India (Mukherji, 2007; Shah, 2007). This argument is based on the premise that regions such as eastern UP, West Bengal, Assam and Bihar depend heavily on diesel power for lifting groundwater, and an increase in price of diesel is likely to raise irrigation costs, significantly reducing farm incomes, as agricultural productivities are already very low there (Kumar et al., 2010).

The empirical work on impact of energy price hikes on irrigated farming, which is based on respondent surveys (see, for instance, Shah, 2007), have little relevance for practical policy formulation in the sense that the perceived 'impacts of price changes' are an outcome of the whole range of changes happening with the farming system, including that on the market front. Such analyses fail to segregate the response of farmer to input price changes, and their subsequent implications for prospects of farming, especially of small and marginal farmers. While farm distress can also be caused by unfavorable changes in output markets, such analysis fails to nullify the effect of changes in output prices (Kumar et al., 2010).

Chapter 7 analyzes how the small and marginal farmers in water-abundant regions respond to a diesel price hike, and assesses the overall impact on the economic prospects of farming. More specifically, it analyzes the actual change in cost of irrigation water for different factors due to rise in diesel prices; the response of diesel well owners to a hike in irrigation costs and its overall impact on economic prospects of farming; and studies the response of water buyers in diesel well commands to price hikes, and the overall impact on economic prospects of their farming.

The recent policy decision of the government of West Bengal to offer heavily subsidized power connections for well irrigation, and to remove the restrictions on issuing permits for drilling new energized wells has not taken cognizance of the situation vis-à-vis arable land and agro-ecology and other socio-economic realities of the state. Chapter 8 analyses the impact of these policies on West Bengal's agriculture, energy economy and water ecosystems. Within agricultural economy, particular focus is on the equity impacts vis-à-vis access to groundwater for irrigation for different socio-economic segments, by assessing how the pump irrigation markets would be influenced by the policy shift.

Many millions of households in India do not have access to 'tap' connections at home. Given the informal nature of the sources and 'services,' the data on actual water use by the households in the communities are absent. The problem is compounded by the lack of clarity on the supply norms for fulfilling multiple water needs of rural population. The sources that are reliable and that can provide adequate quantity of water of sufficient quality to meet various productive and domestic needs seem to be far less than adequate in rural areas, particularly those that are in naturally water-scarce regions. In such regions, the rural poor tend to compromise on their basic needs. Conversely, the communities spend substantial amounts of time and effort in collecting water from distant sources, often of poor quality, to meet requirements of drinking and cooking, domestic uses, livestock uses and kitchen gardens. In both cases, there are undesirable outcomes on health and hygiene, and livelihoods.

A well-designed and effectively implemented 'water supply surveillance' in relation to domestic and productive needs of the community is important input in water supply improvements. The key to designing such a program is information about the adequacy of water supplies and the health and livelihood security risks faced by populations due to lack of it at various levels, to help identify vulnerable areas. There are a range of natural, physical, social, human, economic, financial and institutional factors influencing the vulnerability of the rural population to problems associated with inadequate supply of water for consumption and production needs. They are not captured in the traditional surveillance programs. A composite index to measure the vulnerability of rural households to problems associated with lack of water for domestic and productive needs is presented in Chapter 9.

Likewise, there are thousands of habitations in Maharashtra who face acute shortage of water for domestic and productive uses or have poor access to dependable and safe sources during peak summer months, owing to drying up of 'public water supply sources.' The households in such habitations are vulnerable to problems of health and livelihood associated with water stress during summer months, especially during droughts. In contrast, recent studies indicate that a huge amount of water from large reservoirs scattered all over the state, which are meant for both irrigation and domestic water supplies, remain unutilized toward the end of summer, due to sheer lack of infrastructure for transporting this water to places of demand. On the other hand, in spite of mounting evidence that groundwater-based schemes are highly unsustainable for rural water supply, thousands of villages in the state have gone for such local, village-based schemes, as the Panchayats, the local self-governing institutions, find them easier to manage.

Chapter 10 analyzes the institutional and policy framework for rural water supply in Maharashtra, with particular reference to the 'Sector Reform' initiated by the government in 1999 to see how it influences the performance of rural water supply schemes. For this, the performance of rural water supply sector was reviewed. Then the performance of selected single village and regional water supply schemes with different types of sources – viz., groundwater, river lifting and reservoir – was assessed and compared for effectiveness in management, governance and decentralization. The analysis identifies the most appropriate techno-institutional model for rural water supply, which can ensure sustainability in service delivery. It also illustrates how a 'demand-driven approach' to rural water supply, with an accent on community-based management of the scheme, had seriously compromised on source 'sustainability,' affecting the management performance of the schemes.

With the concept of 'virtual water trade' gaining international attention, it is widely believed that one of the ways to manage water for local and regional water economies in naturally water-scarce regions is to manage water from the global system through food trade (import) (Allan, 1997; Allan, 2003), as food production accounts for the largest share of water amongst all consumptive water uses and the largest demand for water comes from it (Kumar and Singh, 2005).

For several decades, international food trade has been dominated by the West, particularly, the United States, Canada, Australia and some countries in Europe. In the recent past, China, Japan and some agribusiness corporations and billionaires in the Middle East have started participating in international food trade (World

Economic Forum, 2011). Purchase of vast tracts of arable land, which is lying uncultivated, in countries of sub-Saharan Africa, production of crops and their export to the respective countries, are organizing features of this emerging phenomenon. This changing political economy of global agribusiness and food trade has fascinated the academics. Recently, some scholars have noted that attempts by the rising economies of the East to take over the more rooted, agribusiness corporations of the West in international food trade may be part of the effort by these neo-rich countries to grab precious water resources (Sojamo et al., 2012).

The argument that the Western agri-business corporations had established hegemony over global virtual water trade, and that this hegemony over 'international virtual water flow' is being challenged by the rise of southeastern economic powers is highly distorted and has its origin in the incorrect understanding of what determines global agricultural commodity trade (Kumar, 2012). Chapter 11 looks into the real drivers of change in global food trade. For this, it extensively uses the analysis of global data on virtual water flows, through agricultural commodity done by Kumar and Singh (2005). It shows how important it is to not look at food trade from a virtual water lens, and that a correct understanding of the determinants of global agricultural commodity trade would prevent us from developing complacency with respect to food security challenges for water-rich regions.

The last chapter offers a summary of the findings from each one of the chapters, and draws useful suggestions for managing the water-food-energy nexus in developing countries. It illustrates how managing water, energy and food under a scarcity regime requires sound knowledge of statecraft and economic management. The chapter also illustrates the growing 'politics-bureaucracy-academics' nexus in developing countries and discusses how this is leading to pervasive policies on development and use of natural resources, which are causing huge negative impacts on the management of natural resources and their economies.

Notes

1. FAO projections show that feeding a world population of 9.1 billion people in 2050 would require raising overall food production by some 70% between 2005/2007 and 2050. This implies significant increases in the production of several key commodities. Annual cereal production, for instance, would have to grow by almost one billion tons, and meat production by over 200 million tons to a total of 470 million tons in 2050, 72% of which would need to come from the developing countries – up from the 58% today (FAO, 2009).
2. The per capita electricity consumption in India was recorded to be equivalent to 560 kilograms of oil in 2009, and that of China in 2006 was equivalent to 1,695 kilograms of oil (World Bank, 2012).
3. In intensively groundwater irrigated countries like India, the nexus between energy and irrigation is inextricable. In India, around 12.8 million electric pumps with a total of 51.84 gigawatts (GW) of connected load consumed 87.09 billion kilowatt-hours (kWh) of electricity in 2003–2004 (India Energy Portal, n.d.), to irrigate around 55–60 million ha of land (gross). Around 2,400 million liters of diesel is used to power around 6.1 million diesel engines in shallow groundwater areas to irrigate another 25–30 million ha (gross).

4. In the United States, nearly 40% of the water diverted is for thermal power generation, and thermal power accounts for nearly 4% of the total water consumed in various sectors.
5. As far back as 2001, there were around 4 million diesel pump sets in eastern India and around 2.2 million pumps in western and peninsular India (Government of India, 2003).
6. Various pilot studies reveal the poor level of energy efficiency of agricultural pump sets. In an energy audit of electrical pump sets at four field study locations in Haryana, average pump set efficiency was found to be only 21–24% (World Bank, 2001). The study also found that only 2% of the pumps surveyed had efficiency levels above 40%. Phadke et al. (2005) found that a DSM program for replacing inefficient agricultural pumps in Maharashtra would be cost effective by lowering the short-run cost of electricity generation in the state. More recently, an energy audit of a sample of pump sets in Bangalore Rural District in Karnataka was conducted under the Water and Energy Nexus (WEN-EXA) Project of the USAID. The study revealed that 91% of the pumps were operating at the efficiency of less than 30% (Oza, 2007).
7. Households need water for meeting livestock needs, particularly livestock drinking. Rural households, which do not have their own farmland and irrigation sources, prefer water for growing vegetables to meet their domestic needs, as it is essential for the nutritional security of the families. Economically backward rural households, which are not engaged in agriculture, may need water for meeting one or more of the productive water needs for activities such as pottery, fishery, pickle making and duck keeping.
8. This can happen because of two reasons: 1) available water gets reallocated; and 2) the families spend substantial amounts of time and effort to find water to meet the productive needs, which would reduce their ability to work.

References

Adhikari, A. and Bhatia, K. (2010). NREGA Wage Payments: Can We Bank on Banks? Insight. *Economic and Political Weekly, XLV*(1), 30–37.

Aggarwal, P. K., Talukdar, K.K. and Mall, R.K. (2000). Potential Yields of the Rice-Wheat System in the Indo-Gangetic Plains of India. *Rice-Wheat Consortium Paper Series 10. Rice-Wheat Consortium for the Indo-Gangetic Plains* (p. 11). New Delhi, India, and Indian Agricultural Research Institute: New Delhi, India.

Allan, J. A. (1997). *Virtual Water: A Long Term Solution for Water Short Middle Eastern Economies?* Paper presented at the 1997 British Association Festival of Science, University of Leeds, September 9.

Allan, J. A. (2003). Virtual Water Eliminates Water Wars? A Case Study From the Middle East. Virtual Water Trade, Proceedings of the International Expert Meeting on Virtual Water Trade. A. Y. Hoekstra (Ed.), *Value of Water Research Report Series #12*. IHE, Delft, The Netherlands.

Ambasta, P., Vijay Shankar, P. S. and Shah, M. (2008). Two Years of NREGA: The Road Ahead. *Economic and Political Weekly, 43*(8), 41–50.

Byerlee, D., Diao, X. and Jackson, C. (2005). *Agriculture, Rural Development, and Pro-Poor Growth. Country Experiences in the Post-Reform Era.* Synthesis paper for the Operationalizing Pro-Poor Growth in the 1990s Project. Washington, DC: The World Bank.

Datt, G. and Ravallion, M. (1998). Farm productivity and rural poverty in India. *Journal of Development Studies, 34*(4), 62–85.

Earth Policy Institute (2011). Data Center: Climate, Energy, and Transportation. Retrieved from http://www.earth-policy.org/data_center/C23

Evenson, R. E., Carl, E. P. and Mark, W. R. (1999). *Agricultural Research and Productivity Growth in India.* Washington, DC: Research Report 109, International Food Policy Research Institute.

ExxonMobile (2013). The Outlook for Energy: A View to 2040. Retrieved from ExxonMo bil.com/energyoutlook

Food and Agriculture Organization (2000). FAOAQUASTAT database on line on water and agriculture. Retrieved from http://www.fao.org/ag/agl/aglw/aquastat/dbase/

Food and Agriculture Organization (2009). *How to Feed the World 2050.* High Level Expert Forum, Rome, October 12–13.

Government of India (2003). *All India Report on Agricultural Census (2001–02).* Ministry of Agriculture, Government of India: New Delhi.

Grey, D. and Sadoff, C. (2007). Sink or Swim: Water Security for Growth and Development. *Water Policy, 9*(6), 545–571.

Human Development Report (2006). *Beyond Scarcity: Power, Poverty and the Global Water Crisis.* New York: United Nations Development Programme.

India Energy Portal (n.d.). Energy Sectors, Agriculture. Retrieved from http://www.indi aenergyportal.org

International Food Policy Research Institute (2011). *2011 Global Hunger Index: The Challenge of Hunger-Taming Price Spikes and Excessive Food Price Volatility.* Washington, DC: International Food Policy Research Institute, Concern Worldwide and Welthungerhilfe.

Kemper, K. E. (2007). Instruments and Institutions for Groundwater Management. M. Giordano and K. G Villholth (Eds.), *The Agricultural Groundwater Revolution: Opportunities and Threats to Development* (pp. 153–172). CABI: Oxfordshire, UK in association with International Water Management Institute, Colombo, Sri Lanka.

Kumar, M. D. (2007). *Groundwater Management in India: Physical, Institutional and Policy Alternatives.* New Delhi, India: Sage Publications.

Kumar, M. D. (2009). *Water Management in India: What Works, What Doesn't?* New Delhi: India: Gyan Publishing House.

Kumar, M. D. (2010). *Managing Water in River Basins: Hydrology, Economics and Institutions.* New Delhi, India: Oxford University Press.

Kumar, M. D. (2012). Does Corporate Agribusiness Have a Positive Role in Global Food and Water Security? *Water International, 37*(3), 41–45.

Kumar, M. D., Ghosh, S., Patel, A., Singh, O. P. and Ravindranath, R. (2006). Rainwater Harvesting in India: Some Critical Issue for Basin Planning and Research. *Land Use and Water Resources Research, 6,* 1–17.

Kumar, M. D., Narayanamoorthy, A. and Sivamohan, M. V. K. (2010). Pampered Views and Parrot Talks. *The Cause of Well Irrigation in India, Occasional Paper #1,* Hyderabad, India: Institute for Resource Analysis and Policy.

Kumar, M. D., Patel, A., Ravindranath, R. and Singh O. P. (2008). Chasing a Mirage: Water Harvesting and Artificial Recharge in Naturally Water-Scarce Regions. *Economic and Political Weekly, 43*(35), 61–71.

Kumar, M. D., Scott, C. A. and Singh O. P. (2013b). Can India Raise Agricultural Productivity While Reducing Groundwater and Energy Use? *International Journal of Water Resources Development, 29*(4), 1–17. Routledge, iFirst article. Retrieved from http://www.tandfonline.com/eprint/BWD2sAaiTFkIG4kmxUwr/full

Kumar, M. D. and Singh O. P. (2005). Virtual Water in Global Food and Water Policy Making: Is There a Need for Rethinking? *Water Resources Management, 19*(6), 759–789.

Kumar, M. D., Sivamohan, M. V. K. and Bassi, N. (Eds.). (2013a). *Water Management, Food Security and Sustainable Agriculture in Developing Economies.* Oxon, United Kingdom: Routledge.

Kumar, M. D., Sivamohan, M. V. K. and Narayanamoorthy, A. (2012). The Food Security Challenge of the Food-Land-Water Nexus in India. *Food Security, 4*(4), 539–556.

Ladha, J. K., Fischer, K. S., Hossain, M., Hobbs, P. R. and Hardy, B. (2000). Improving the Productivity and Sustainability of Rice-Wheat Systems of the Indo-Gangetic Plains: A Synthesis of NARS-IRRI Partnership Research. Discussion Paper No. 40 (pp. 1–31). Philippines: International Rice Research Institute.

Mukherji, A. (2007). The Energy-Irrigation Nexus and Its Impact on Groundwater Markets in Eastern Indo-Gangetic Basin: Evidence From West Bengal, India. *Energy Policy, 35*(12), 6413–6430.

Mukherji, A. (2008). *The Paradox of Groundwater Scarcity Amidst Plenty and Its Implications for Food Security and Poverty Alleviation in West Bengal, India: What Can Be Done to Ameliorate the Crisis?* Paper presented at 9th Annual Global Development Network Conference, Brisbane, Australia, January 29–31.

Mukherji, A., Shah, T. and Banerjee, P. (2012). Kick-Starting a Second Green Revolution in Bengal: Commentary. *Economic and Political Weekly, 47*(18), 27–30.

National Council of Applied Economic Research/Public Interest Foundation (2009). NCAER-PIF Study on Evaluating Performance of National Rural Employment Guarantee Act. New Delhi, India: National Council of Applied Economic Research.

Oza, A. (2007). *Irrigation: Achievements and Challenges, India Infrastructure Report (2007)*. New Delhi, India: Oxford University Press.

Pathak, H., Ladha, J. K., Aggarwal, P. K., Peng, S., Das, S., Singh, Y., . . . Gupta, R. K. (2003). Trends of Climatic Potential and On-Farm Yields of Rice and Wheat in the Indo-Gangetic Plains. *Fields Crops Research, 80*(3), 223–234.

Phadke A., Sathaye J. and Padmanabhan, S. (2005). Economic Benefits of Reducing Maharashtra's Electricity Shortage Through End-Use Efficiency Improvement. *LBNL Report 57053*. Berkeley, CA: Lawrence Berkeley National Laboratory.

Rao, V. V. K. (2006). Hydropower Potential in the North East: Potential and Harnessing Analysis. Background Paper #6. Prepared as an input for the study *Development and Growth in Northeast India: The Natural Resources, Water, and Environment Nexus*.

Saleth, R. M. (1997). Power Tariff Policy for Groundwater Regulation: Efficiency, Equity and Sustainability. *Artha Vijnana, 39*(3), 312–322.

Shah, T. (1993). *Groundwater Markets and Irrigation Development in India: Political Economy and Practical Policy*. New Delhi, India: Oxford University Press.

Shah, T. (2001). Wells and Welfare in Ganga Basin: Public Policy and Private Initiative in Eastern Uttar Pradesh, India. *Research Report 54*. Colombo, Sri Lanka: International Water Management Institute.

Shah, T. (2007). Crop Per Drop of Diesel? Energy Squeeze on India's Smallholder Irrigation. *Economic and Political Weekly, XLII*(39), 4002–4009.

Shah, T., Gulati, A., Hemant, P., Shreedhar, G. and Jain, R. C. (2009). Secret of Gujarat's Agrarian Miracle after 2000. *Economic and Political Weekly*, 44(52), 45-55.

Shah, T., Roy, A. D., Qureshi, A. S. and Wang, J. (2003). Sustaining Asia's Groundwater Boom: An Overview of Issues and Evidence. *Natural Resources Forum, 27*(2), 130–141.

Shah, T., Scott, C. A., Kishore, A. and Sharma, A. (2004). Energy Irrigation Nexus in South Asia: Improving Groundwater Conservation and Power Sector Viability. *Research Report No. 70*. Colombo, Sri Lanka: International Water Management Institute.

Shah, T., Scott, C., Kishore, A. and Sharma, A. (2007). Energy-Irrigation Nexus in South Asia: Improving Groundwater Conservation and Power Sector Viability. In M. Giordano and K. Villholth (Eds.), *The Agricultural Groundwater Revolution: Opportunities and Threats to Development* (pp. 211–242). Wallingford, UK: CABI Publishing.

Sikka, A. K. and Bhatnagar, P. R. (2006). Realizing the Potential: Using Pumps to Enhance Productivity in the Eastern Indo-Gangetic Plains. In B. R. Sharma, K. G. Villholth and K. D. Sharma (Eds.), *Groundwater Research and Management: Integrating Science Into Management Decisions, Groundwater Governance in Asia Series-1*. Colombo, Sri Lanka: IWMI.

Singh, A. (2009). *A Policy for Improving Efficiency of Agriculture Pump Sets in India: Drivers, Barriers and Indicators*. Kanpur: International Support for Domestic Action, Indian Institute of Technology.

Sojamo, S., Martin, K., Warner, J. and Allan, A. J. (2012). Virtual Water Hegemony: The Role of Agribusiness in Global Water Governance. *Water International, 37*(2), 69–182.

Svendsen, M., Ewing, M. and Msangi, S. (2009). *Measuring Irrigation Performance in Africa*. IFPRI Discussion Paper No. 894. Washington, DC: IFPRI.

The 2030 Water Resources Group (2009). *Charting Our Water Future-Economic Frameworks to Inform Decision Making*. Author.

The United States Department of Energy (2005, August). *Power Plant Water Usage and Loss Study*. Report prepared for National Energy Technology Laboratory. The United States Department of Energy.

Tiwari et al. (2011). MNREGA for Environmental Service Enhancement and Vulnerability Reduction: Rapid Appraisal in Chitradurga District, Karnataka. *Economic and Political Weekly, 46*(20), 39–47.

Vanaik, A. and Siddhartha (2008). Bank Payments: End of Corruption in NREGA? *Economic and Political Weekly, 43*(17), 33–39.

World Bank (2001). *India: Power Supply to Agriculture*. South Asia Region: Author.

World Bank (2010). Deep Well Prudence: Towards Pragmatic Action for Addressing Groundwater Over-Exploitation in India. Washington, DC: Author.

World Bank (2012). *The World Development Report (2012)*. Washington, DC: Author.

World Economic Forum (2011). *Water Security: The Water-Food-Energy-Climate Nexus*. Washington/Covelo/London: The World Economic Forum Water Initiative, Island Press.

2 Unraveling Gujarat's agricultural growth story

M. Dinesh Kumar, A. Narayanamoorthy, O. P. Singh, M. V. K. Sivamohan and Nitin Bassi

Gujarat's agricultural growth story

The poor growth in agriculture during recent years has been a matter of grave concern for the policymakers in India (Planning Commission, 2008; Bhalla and Singh, 2009). The blame has been on poor natural resource conservation policies; poor design of subsidies; inadequate investments in irrigation; inefficient pricing of water, electricity and other inputs for crop production; and poor agricultural pricing policies and regulations such as bans on interstate trading of agricultural produce, particularly cereals. But, least is written about how poor management of water economy is causing long-term effects on Indian agriculture, particularly in regions that are historically agriculturally prosperous.

In this backdrop, Gujarat's agriculture sector has been in the limelight for the 'high growth' it has recorded in the early years of the new millennium (see Gulati et al., 2009; Shah et al., 2009). After Gulati et al. (2009), the state has clocked an impressive growth rate of 9.6% in the sector. The key state interventions, which have potential implications for agriculture in the state, are as follows: improved quality of power supply in agriculture; large-scale water transfers from land-scarce and water-abundant south Gujarat to land-rich and water-scarce north Gujarat; decision to meter new agro wells; setting up of the Gujarat Green Revolution Company for promoting micro irrigation adoption; and decentralized water harvesting.

But some researchers have attributed this phenomenal growth to selected policies adopted by the state in the water and electricity sector. For instance, the most recent article by Shah et al. (2009) argued that agricultural growth in the state has mainly come from three factors: increase in gross cropped area (GCA); increase in productivity (yield per ha); and increase in farm gate price. They further argued on the basis of very limited data for a short time period of 7 years that north Gujarat, Saurashtra and Kachchh have mainly contributed to the GCA expansion. They assessed that the irrigated areas of Saurashtra, Kachchh and north Gujarat generate much higher wealth as compared to the canal irrigated areas of south Gujarat from every unit of water used for crop production. This was attributed to lacs of small water harvesting structures, which came up and were claimed to have been built through community participation in these regions. They in turn were said to provide 'supplementary irrigation' to crops and a highly productive well irrigation, which was supported by an 8-hour uninterrupted power supply to farmers under a new scheme called 'Jyotigram Yojna.' Some others showed that in the post-liberalization period (1990–93 to 2003–06), agricultural growth in Gujarat has been higher than in the pre-liberalization period, while in the country as a whole it decelerated (Bhalla and Singh, 2009).

This 'growth story' needs further examination. Such an exercise can provide clues on framing agricultural policies for the country as a whole, given the fact that Gujarat is a representative part of India because it displays significant regional differences in socio-economic conditions, the agro-ecology (soils, climate and topography), the water resources endowment and the condition of rural infrastructure. We begin with the proposition that 5- to 6-year duration is too short a timeframe for one to make any assessment of agricultural growth in a state like Gujarat, which has a highly fragile agro-ecological system, that the real miracle growth had occurred in the previous decade (from 1988 to 1998) and that the impressive growth displayed recently is nothing more than a recovery of the sector after a major dip in outputs owing to severe droughts.

In this chapter, we attempt a reality check on the much-believed miracle growth in Gujarat's agricultural production by looking at the gross value of the outputs from agriculture over a reasonably long period of time. Subsequently, the key sub sectors that have contributed to this growth are identified; and the trends in cropped area, yield and total production are systematically examined. Further, the factors that might have actually changed the agriculture growth scenario in the state are identified. In order to identify this, the factors that have the potential to be the drivers of agricultural output growth in the region are identified, on the grounds of the physical, socio-economic, environmental, institutional and policy contexts. In other words, an assessment of the real constraints to agriculture growth is made. Further, we examine how these constraints have been tackled over a period of time.

Agricultural growth: Some theoretical perspectives

Agricultural growth in any region can occur because of: 1) growth in crop output; 2) increase in value of the given output and 3) diversification of agriculture toward high-valued crops and livestock products (Bhalla and Singh, 2009). Here, the growth in output can result from three major phenomena. First, the output of a crop can increase due to a variety of reasons, including crop technology adoption, irrigation supplement to rain-fed crop, precision irrigation, availability of adequate soil moisture due to rains and better soil nutrient management or increase in area under the given crop. Second, the value of the given output in the market can increase due to changes in the demand–supply situation, which is particularly important in the case of noncereal crops and perishable products such as fruits and vegetables, and where the sufficient infrastructure for storage is either absent or economically unviable. Third, the farmers can shift to high-valued crops or livestock, which give higher returns from unit of land and unit of livestock, respectively. Such a shift can often be subject to high crop risk or market risk (Kumar and Amarasinghe, 2009). But availability of good credit facilities, marketing infrastructure, research and extension services and technical inputs can speed up this process.

Hence, several factors can drive agricultural growth, viz., environmental factors; institutional factors, including those that are market related (Bhalla and Singh, 2009; Gulati, 2002); policy factors; infrastructural factors (Bhalla and Singh, 2009; Shah et al., 2009) and science/technology-related factors (Bhalla and Singh, 2009). Some of the environmental factors here are: changes in precipitation, changes in atmospheric temperature, humidity, wind speed and sunshine (exogenous) and changes in soil moisture regime and soil nutrient regime (endogenous). The infrastructure-related factors are: the presence of irrigation facility, presence of roads for transport, presence of storage and market infrastructure and precision irrigation technology. That said, it is important to consider that creation of both irrigation infrastructure (wells, reservoirs – both small and large – canals and pumps) and installation of precision water application technologies will have their effects only if the resource availability situation is good or does not get altered. In the face of resource depletion (like reduced inflows into irrigation reservoirs or groundwater depletion, showing up in declining well yields), the potential benefits of extended irrigation infrastructure in the form of expansion in irrigated area cannot be derived.

The input price policy can be one that encourages efficient use or one that encourages wasteful use of input resources such as water, fertilizers and pesticides. For the first one to happen, the price has to be raised to reflect the value of the resource, and for the second one, the price has to be lowered or input subsidy raised. While the positive impact of the first one will be both long term and short term (Pearce and Warford, 1993), that of the second one will be short term. Conversely, the negative effects of the first measure, if any, would be rather short lived, and that of second measure would be long term. The best example is electricity pricing for groundwater pumping.

Institutional factors such as property rights in land and water will have long-term impacts on the equity, efficiency and sustainability of resource use (Pearce and Warford, 1993; Thobani, 1997) – here use of land and water. Similarly, good extension services will have both short- and long-term impacts on yield, by encouraging farmers to adopt better crop varieties, or better input use technologies or better agronomic practices. Produce price regulations, particularly through enforcement of minimum support prices, will have a significant impact on allocation of land for a particular crop (Bhalla and Singh, 2009). Also, agricultural trade can have severe effects on the market value of crops, which the trade affects. But, in the long run, the volume of production of that crop itself can change. In which direction the change takes place depends on whether the trade policy encourages import or export of the commodity in question, and the comparative advantage of the region in question in terms of producing that crop.

Lastly, the advancing science and technology can have far-reaching consequences, and can often overcome the constraints induced by the environmental factors. This has been adequately demonstrated in India, which experienced the impacts of high-yielding varieties of major cereal crops, brought in by the Green Revolution. Introduction of a new high-yielding variety or a drought-resistant variety can have both short-term and long-term effects on crop productivity, depending on the environmental factors. In a region that is highly vulnerable to droughts, a drought-resistant variety of a dominant crop will have a significant impact on both the area under the crop and the yield of the crop, if significant drought-proofing measures are not in place. But, in the face of deteriorating soil quality (or primary productivity of soils), the potential benefits from high-yielding varieties cannot be derived. This is called 'technology fatigue.'

All these lead us to the point that the drivers of growth in agricultural could be too many, and the final outcome of introduction of these drivers would be a result of the interplay of different drivers. A policy to boost groundwater irrigation will have the desired effects on irrigated areas, unless sufficient groundwater is available for exploitation or sufficient arable land is available for expansion of cropped area. To sum up, it would be meaningless to make linear or even unidirectional projections of the impacts of one set of interventions, be it policy related, institutional, market, technology related or infrastructure related, without knowing the interactions amongst various actors.

Gujarat's growth: Long-term or short-term?

Gujarat witnessed one of the worst droughts of the last century for 3 consecutive years from 1985 to 1987 (Bhatia, 1992). Drinking water had to be transported to Rajkot by train, the cost of which was more than the cost of desalination of seawater at that point in time. It is also known that the state witnessed another severe drought for 2 years from 1999 to 2000. A graphical representation of the value of agricultural outputs in Gujarat for the period from 1980–81 to 2005–06 is given in Figure 2.1. It can be seen that during these years, as data on value of agricultural output from the state shows, the agricultural outputs fell remarkably. The fall was to the tune of 56% from 1984 to 1987, and 30% from 1998 to 2000. The effect of the 1987 drought is for the crop year of 1987–88 and that of 2000 is for the crop year of 2000–01. Hence any growth projections that consider these years (i.e., 1987–88 and 2000–01) as the base year can give a misleading picture of the growth scenario.

In order to study the agricultural growth in Gujarat, the data for 11 years from 1988–89 (corresponding to the good rainfall year of 1988) to 1998–99 are taken. This was compared against the growth figures for the period from 1998–99 (corresponding to the normal year of 1998) to 2009–10.[1] In both the cases, the constant prices as well as the current prices were considered.

Our analysis shows that agricultural growth, based on value of the outputs at current prices during the 11-year period, which included the initial years of economic liberalization, was dramatic, and the annual compounded growth rate clocked a figure of 20.8%. Prior to that, agriculture in Gujarat did not grow much, from 1980–81 till 1987–88 due to several factors, most important of which was the severe drought of 1985–87. Also, the growth during the subsequent period

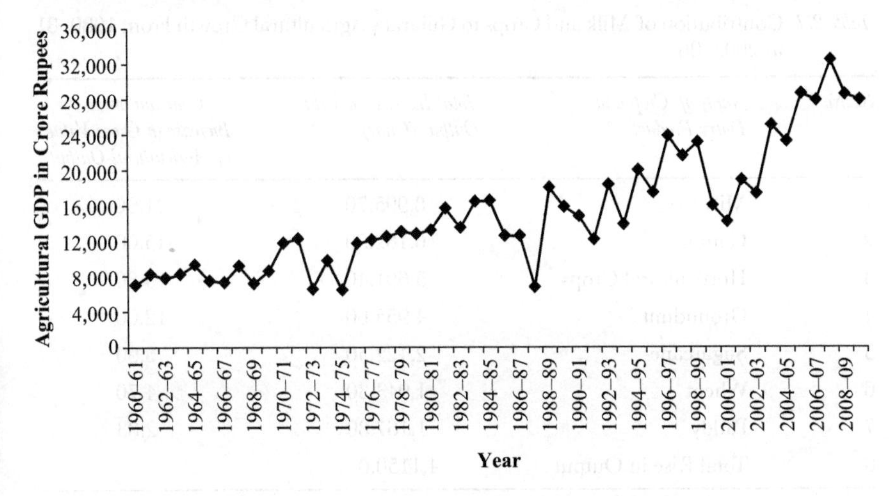

Figure 2.1 Agricultural GDP in Gujarat in Crore Rupees at Constant Prices (1960–61 to 2009–10)

(i.e., 1998–99 to 2009–10) was a meager 7.4%. The growth in real terms in the first case was, however, 2.5%, whereas in the second case it was only 1.7%. Hence, we can safely argue that the real 'growth' in Gujarat's agriculture occurred during the decade from 1988–89 to 1998–99. These analyses question the validity of the recent argument made by Gulati et al. (2009) and Shah et al. (2009) that Gujarat witnessed 'miracle growth' since 2000–01.

Where has the growth come from?

It is important to find out the crops that have actually contributed to this growth. This is crucial to identify the factors that are driving this growth. For this, the changes in gross value of total agricultural outputs (at current prices) during the period from 1980–81 to 2006–07 were analyzed using data from Central Statistical Organization, and compared against that of individual produce. Our analysis shows that the increase in gross value of agricultural outputs in the state was in the tune of 41,150 crore rupees. Five major agricultural products, which have contributed to the growth, are: milk, cotton, horticultural crops, groundnut and sugarcane. Wheat and paddy take 6th and 7th place (Table 2.1). This clearly shows that dairy production remains the frontrunner in Gujarat's agricultural growth parade.

Now, one can also argue that the recent growth is the result of policy measures adopted by the state. For instance, it is vehemently argued that introduction of Bt. cotton, which has caught like 'wildfire' in Gujarat, had mainly contributed to the growth. Further, it is argued that area expansion in wheat is occurring as a result of improved groundwater situation owing to extensive, decentralized and small-scale water harvesting in Saurashtra, Kachchh and north Gujarat, has driven the growth (Shah et al., 2009).

Table 2.1 Contribution of Milk and Crops to Gujarat's Agricultural Growth From 1980–81 to 2005–06

Sr. No	*Name of Crop and Dairy Product*	*Total Increase in Value Output (Crore)*	*% Contribution to the Increase in Gross Value of Agricultural Output*
1	Milk	8,995.70	21.90
2	Cotton	6,162.90	15.00
3	Horticultural Crops	5,691.40	13.80
4	Groundnut	4,955.60	12.00
5	Sugarcane	2,422.50	5.90
6	Wheat	1,943.30	4.70
7	Paddy	1,167.80	2.83
8	Total Rise in Output	4,1150.0	

Source: Authors' own estimates based on GOI, 1996, 2004, 2008

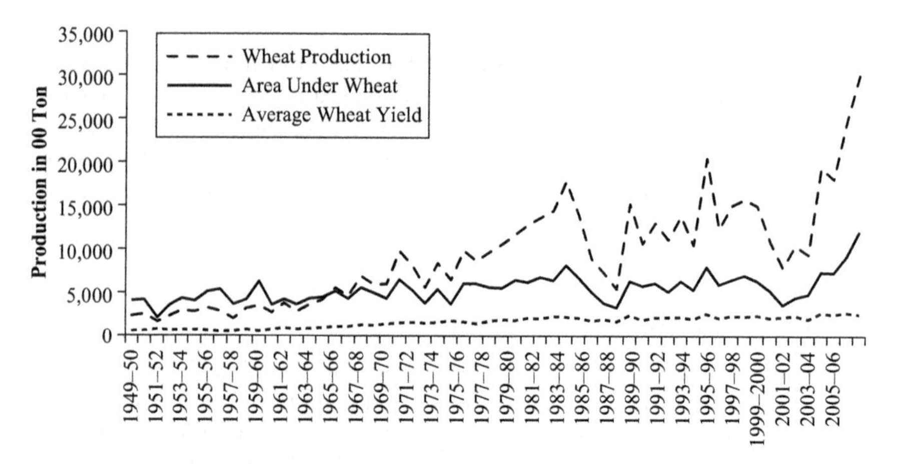

Figure 2.2 Wheat Production Trends in Gujarat (1949–50 to 2006–07)

But analysis of historical data (Crop, area and outputs of major crops in Gujarat, 1949–2006, Government of Gujarat) shows a totally different trend. The real, dramatic and steady growth in wheat production in Gujarat occurred from 1949 onward, with improvements coming from expansion in irrigated wheat replacing rain-fed wheat, and then from the 1970s mainly as the result of adoption of high-yielding varieties. But the yield effect almost disappeared after 1985, with no remarkable change in average crop yield, and further increase in production came from increase in area under the crop (Figure 2.2). But what is more important is the fact that the production has become highly erratic, with sharp declines in production during drought years (i.e., from 1985–87 and then from 1999–2000). There is also a perfect correlation between production and area under the crop.

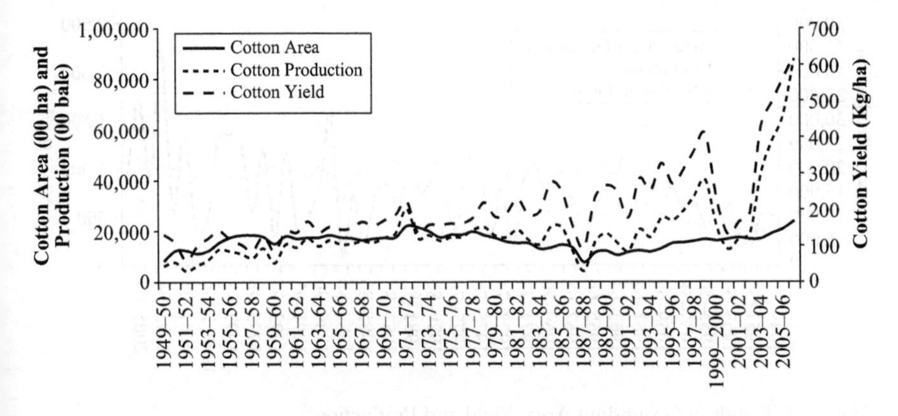

Figure 2.3 Trends in Cotton Area, Yield and Production in Gujarat

Now let us examine what has happened to cotton, a major crop for Gujarat even today, accounting for nearly 13.5% of the total value of the agricultural output in the state. Over the 60-year term, the cotton production in the state has steadily increased 14-fold, and most of this increase came from yield effect and not so much from the area expansion, as is evident from Figure 2.3.

The average yield of cotton in the state has been steadily increasing from a mere 130kg/ha in 1949–50 to 624kg/ha in 2006–07. This could be due to two important factors: 1) replacement of rain-fed cotton by irrigated cotton, and 2) greater use of high-yielding varieties of cotton. Though there has been a marked and consistent increase in area under cotton during 1994 and 2006, this did not get translated into production gain, and there was a sharp decline in yield during the drought years.

Another important crop, which has the potential to turn around the agrarian scenario of Gujarat is groundnut, Saurashtra being known as the 'groundnut bowl' of India. The area under groundnut, which is the most dominant kharif crop of the Saurashtra region, has been hovering around 2.0 m. ha during the past 3 decades or so, after a slow decline from a peak of 2.3 m. ha in the early 1960s. An exception is the shrinking of area that occurred during 1987–88, the third year of the most severe drought of the century. This is quite understandable, as the farmers in the region were facing extreme water shortages after 2 years of drought and did not want to take a major risk.

But what is more striking is the fluctuations in crop output, which can only be explained by the inter-annual yield fluctuations. During virtually every drought, the yield went down drastically (1985–87, 1993, 1999 and 2000) touching the lowest of 203 quintals per ha in 1987. So, the major determinant controlling groundwater production in Saurashtra is the yield, which depends fully on monsoon. The 'recharge movement' started in Saurashtra in 1988, comprising recharging of dug wells by individual farmers. But neither this nor the decentralized water harvesting initiative launched in 1998 by the government of Gujarat (IRMA/UNICEF, 2001) seem to have helped protect the groundnut crop during the drought years of 1999 and 2000. In fact, the yield fluctuation is even severe after 1988, and has become a regular phenomenon (Figure 2.4).

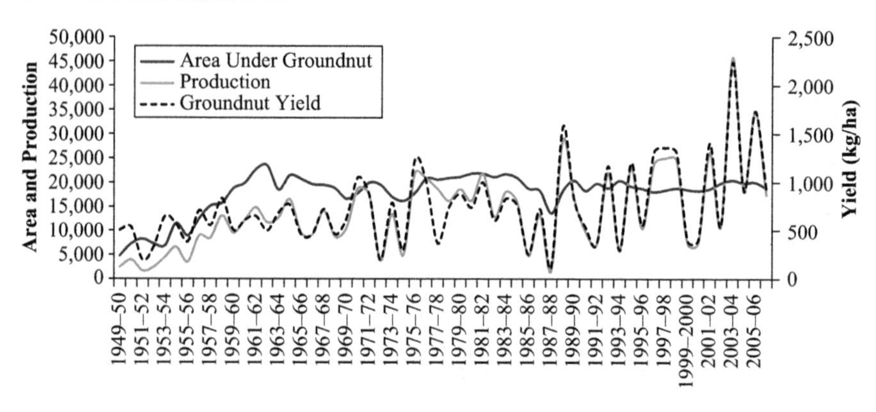

Figure 2.4 Trends in Groundnut Area, Yield and Production

The analysis of area-yield-production trends in three distinct crops of Gujarat reveals the following important points about agricultural production in the state. First, the production of crops, which have a rain-fed component, is highly vulnerable to droughts, with the droughts impacting on the crop yield. The impact of drought on the cropped area is not perceptible. In the case of winter crops, the drought hits the production as a result of farmers reducing the area under the crop in the wake of poor availability of water from the wells or from the surface irrigation systems. It appears quite clearly that the cotton yield, and therefore production, is now becoming highly susceptible to monsoons unlike in the past. The reason is production of irrigated high-yielding cotton varieties, which have replaced low-yielding rain-fed varieties, is heavily dependent on the availability of water not only from the rains but in the aquifers and surface reservoirs. Since, during droughts, both groundwater and surface water availability decline sharply, either the crop fails or the yield is severely affected.

Now, the most important farming enterprise in Gujarat, in terms of contribution to agricultural growth, is dairying. Analysis of data for 15 years (all that is available) shows that it has grown consistently and at a fast rate. The annual compounded growth rate (at current prices) was estimated to be 11.7%. This is a very high growth. What is most important is the fact that this is the only farm produce in the state that had not suffered any setback during the drought years. While the milk output showed a minor decline during 2002–03, this was attributed to very low paddy and wheat area during the year, which might have affected the animal fodder availability. But researchers have ignored this aspect of Gujarat's agrarian economy. It appears that dairying is still the most favorite of Gujarat's farmers, accounting for more than 21% of the wealth generated from agriculture. This remarkable achievement can be attributed to the vibrant dairy institutions in north and south Gujarat, which procure milk from farmers and pay remunerative prices because of the presence of a good processing and marketing infrastructure (Figure 2.5).

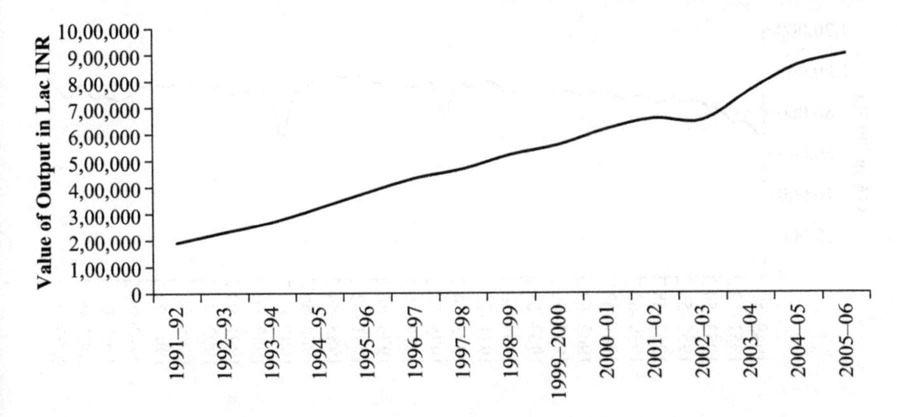

Figure 2.5 Trends in Value of Milk Output in Gujarat

What is happening to the gross cropped area?

Having examined the trends for three important crops and milk, it is now necessary to gauge the changes at the aggregate level, including all the crops grown in the state. The aim is to see the nature of growth and the type of factors that might have influenced the growth. For this, we have examined the trend in gross cropped area over a substantially long period of time, i.e., 58 years. The analysis of data (Figure 2.6) for the period from 1949–50 to 2006–07 (Government of India, 1996, 2004, 2008) shows that there was a negligible long-term growth in gross cropped area.

This is equal to adding nearly 560 ha of land to the cropped area every year. This is statistically insignificant for a state that has a total cultivated area of 10 m. ha. A closer look at the data for different time periods shows that the GCA peaked in 1983–84 (9.949 m. ha), and started declining during droughts and then slowly recovered. A second major depression was observed during the drought of 1999 and 2000. But the recovery after that was never substantial enough to attain the original peak GCA in spite of consecutive wet years. The only exception was a sudden increase in GCA noticed in 2006–07, with the area jumping from 9.3 m. ha to 10.126 m. ha.

So, it seems that till 1984, several factors including expansion in public irrigation schemes, massive rural electrification and the institutional measures for encouraging private investment for groundwater irrigation such as heavy subsidies for electricity in the farm sector, the institutional financing for well development and subsidies for well drilling and installing pumps have driven major growth in cultivated area in the state, through increase in cropping intensity with the help of irrigation. Tube well irrigation has seen an explosion in the alluvial areas of the state, whereas in the hard rock areas of Saurashtra and Kachchh, energized open wells have helped expand groundwater irrigation.

However, the GCA stagnated afterward, with occasional drops encountered during meteorological droughts. While the changes in electricity tariff policy had

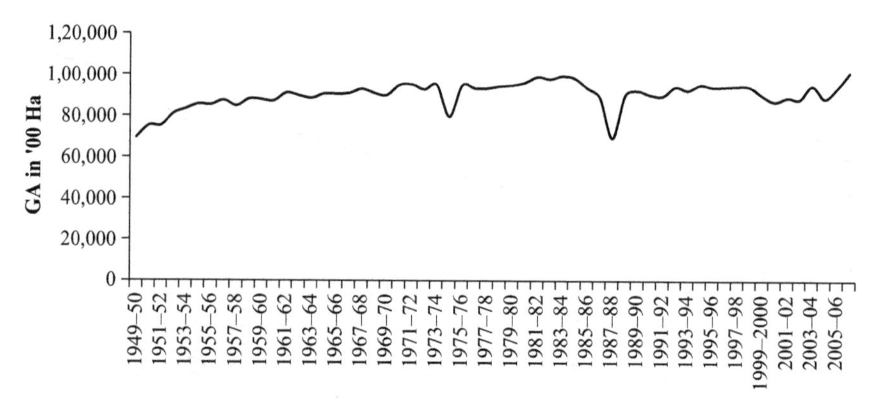

Figure 2.6 Trends in GCA in Gujarat

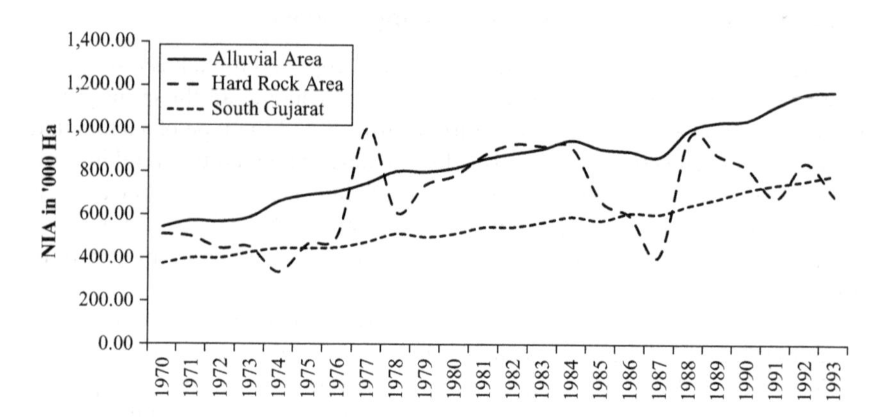

Figure 2.7 Trends in Net Irrigated Area Under Three Distinct Groundwater Environments

been catalytic to greater exploitation of groundwater, following the removal of electricity meters from farms and introduction of flat rate pricing, the constraints induced by the limited water resources, especially groundwater resources in many parts of the state had inhibited further expansion in cropped area. This is more so for hard rock areas underlying the entire Saurashtra region and Sabarkantha district of north Gujarat, which have very meager groundwater stock, and the only source of replenishment of aquifers is the rainfall. As Figure 2.7 shows, the net irrigated area from all sources started declining sharply during droughts, remarkably affecting the GCA. Further, even many good rainfall years do not seem to cause recovery, and the net irrigated area in these hard rock areas seem to be descending after 1988. But mining of the available 'groundwater stock' in alluvial aquifers using tube wells had facilitated conversion of some of the rain-fed crops into irrigated crops in north Gujarat, as is evident from Figure 2.7, which shows consistent increase in net irrigated area in that region. But many farms that have lost direct access to groundwater due to steep rise in cost of well irrigation had reduced their irrigated area.

What can change the agricultural future of Gujarat?

At the aggregate level, Gujarat is one of the water-scarce regions of India with the per capita renewable water resource availability falling far below the 1,700 cubic meter per annum mark.

But there is sharp variation in water resources endowment of the state, with the mean of average annual rainfall varying from 350 mm in Kachchh to 2,000 mm in Valsad in the south (IRMA/UNICEF, 2001). The variability in rainfall also increases sharply with lowest variability in the high rainfall regions in the south and southeastern parts to the highest variability found in Kachchh, followed by Saurashtra and north Gujarat (Figure 2.8). The entire south Gujarat receives moderate to high rainfall varying from 900 mm to 2,000 mm. Saurashtra has an average mean annual rainfall of around 550 mm, and north Gujarat has rainfall varying from 900 mm in the eastern parts to around 400 mm in the western parts. The per capita renewable water resources is 1,832 cubic meters per annum in south Gujarat, 427 cubic meters per annum in north Gujarat, 734 cubic meters per annum in Saurashtra and 875 cubic meters per annum in Kachchh (Figure 2.9). It may be mentioned here that the relatively high renewable water resource in Kachchh, in comparison to north Gujarat, is by virtue of the low population density in the region (IRMA/UNICEF, 2001; Kumar, 2002).

Most of Gujarat's surface water resources are concentrated in south Gujarat, with many perennial rivers such as Mahi, Narmada, Tapi, Karjan and Damanganga, and carry huge amount of flows annually. Due to low variability in rainfall, the variability in annual stream flows is also low, increasing the dependability.

In contrast to this, north Gujarat, Saurashtra and Kachchh have very poor surface water endowment, and the rivers and rivulets there have only limited seasonal flows. Due to high interannual variability in rainfall, the stream flows also

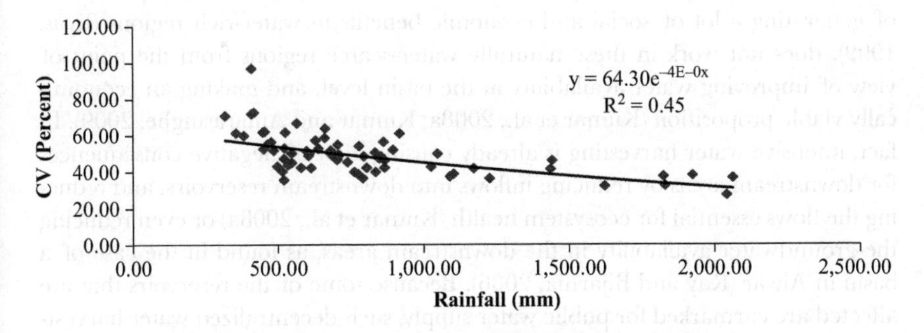

Figure 2.8 Correlation Between Rainfall and Coefficient of Variation

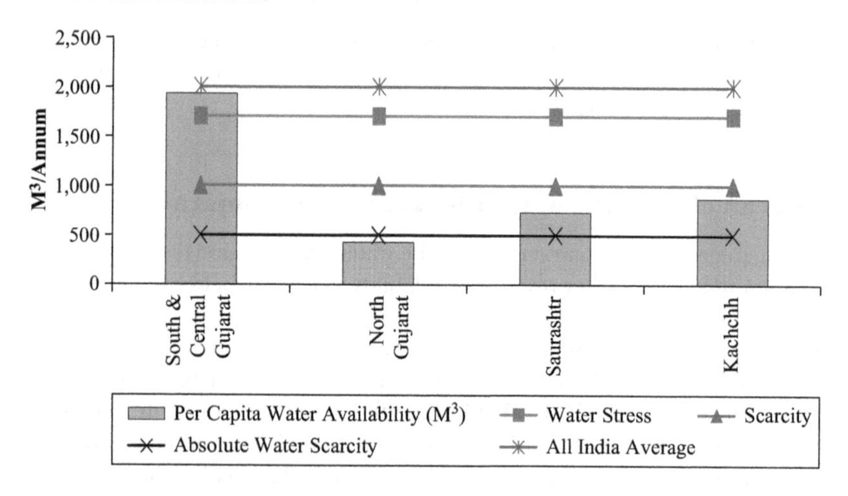

Figure 2.9 Freshwater Availability in Gujarat by Region

vary remarkably from year to year. The numerous major and medium irrigation schemes built in these three regions during the 1960s, 1970s and 1980s impound the monsoon runoff from around 91 basins in Saurashtra, around 100 rivulets in Kachchh and a few small and big river basins in north Gujarat (viz., Sabarmati, Banas, Rupen and Saraswati), and are overdesigned (Kumar, 2002). Because of this, there is hardly any outflow from these reservoirs even in good rainfall years. From a macro perspective, several of the small and large basins in these regions have no residual catchment upstream of these reservoirs. Also, they have negligible residual catchments downstream. On the other hand, groundwater resources in the water-scarce arid and semi-arid regions of Gujarat are already 'overexploited' or are on the verge of it. Further exploitation of groundwater for expanding irrigation is not possible in any of these regions. On the contrary, during droughts, irrigated agriculture can drop sharply, as recharge to aquifers declines while irrigation water requirement goes up. This is one reason why the agricultural outputs fall more sharply these days in the event of droughts, making the growth very erratic.

Therefore, water harvesting, which otherwise is an innovative concept capable of generating a lot of social and economic benefits in water-rich regions (Ilyas, 1999), does not work in these naturally water-scarce regions from the point of view of improving water availability at the basin level, and making an economically viable proposition (Kumar et al., 2008a; Kumar and Amarasinghe, 2009). In fact, intensive water harvesting is already causing a lot of negative consequences for downstream areas by reducing inflows into downstream reservoirs, and reducing the flows essential for ecosystem health (Kumar et al., 2008a) or even reducing the groundwater availability in the downstream areas, as found in the case of a basin in Alwar (Ray and Bijarnia, 2006). Because some of the reservoirs that are affected are earmarked for public water supply, such decentralized water harvesting activities often lead to conflicts (Kumar et al., 2008a). One of the many factors that makes small water harvesting structures economically unviable in naturally water-scarce regions is that the water impounded by these small reservoirs is never diverted for beneficial uses, and instead gets evaporated (Kumar et al., 2006).

At the same time, the still unutilized surface water in south Gujarat basins can be transferred to the water-scarce regions of north Gujarat, Saurashtra and Kachchh, which still have a lot of arable land that can be brought under irrigated production, though the economic viability of the same need to be ascertained. Ranade and Kumar (2004) had argued for transfer of surplus water from the Narmada for recharging the alluvial aquifers in north Gujarat, using gravity recharge by spreading water in the fields, and the initial analysis had shown that this is economically viable. But pricing this water appropriately and introducing electricity tariff reform is very crucial to make sure that the full economic value of the water is realized in this use. In view of this, the recent initiative to transfer water from water-abundant south Gujarat to north Gujarat through Narmada Main Canal is a great step forward. This would not only rejuvenate the ailing agrarian economy of north Gujarat, but would also improve the sustainability of groundwater resources in the region, which is the region's lifeline.

Another source of agricultural growth in Gujarat is crop water productivity improvements, especially in the water-scarce regions of north Gujarat, Saurashtra and Kachchh. The aim will have to be raising the returns from every unit of water used up in agricultural production. Different regions of Gujarat put together around 40 different crops, many of which are amenable to micro irrigation systems such as drips and sprinklers. Some of them are: cotton; castor; groundnut; potato; vegetables such as brinjal and chili; banana; sugarcane and fruit crops such as lemon, pomegranate, gooseberry and mango. Most of these crops, except mango, banana and sugarcane, are grown extensively in the water-scarce regions of the state. According to Kumar and van Dam (2013), these systems can actually

Table 2.2 Impact of Micro Irrigation Technologies on Crop (Applied) Water Productivity on Selected Crops in North Gujarat

Name of Crop	*Crop (Applied) Water Productivity (Rs./m³)*	
	Before adoption of WST	*After adoption of WST*
Potato (micro sprinklers)	7.04	17.99
Cluster Bean (micro sprinklers)	7.68	15.77
Pomegranate (drips)	NA	41.37
Groundnut (micro sprinklers)	4.13	9.36
Cotton (drips)	10.32	18.81
Chili (drip)	34.87	148.1

Source: Authors' own analysis based on primary data, 2009

result in real water saving when used for row crops, in semi-arid and arid climates with deep groundwater table. As an outcome of a project initiated by IWMI called 'North Gujarat Groundwater Initiatives,' many thousands of farmers in the region have adopted MI systems for row crops, accompanied by shifts in cropping patterns to highly water-efficient crops such as pomegranate, chili, lemon, potato and vegetables.[2] A survey of 114 farmers showed remarkable increase in crop water productivity due to this (see Table 2.2). The average farm income rose up by Rs. 99,442 per annum, and was most significant in the case of orchard growers.

What has actually driven the recent 'trend'?

What clearly emerges from the analysis is that the real growth in agricultural production has occurred during 1988–89 to 1998–99. The growth rate was not only high but also steady. Another important fact vis-à-vis growth is that the sectors, which have mainly contributed to this growth, are milk, cotton, fruits and vegetables, sugarcane and groundnut, and wheat is only the last among the six. The 'growth' observed in the recent past (from 2002 onward) is nothing but a good recovery from a major dip in production that occurred after the droughts of 1999 and 2000.

Two important factors have contributed to this recovery: 1) the occurrence of four successful monsoons in the state after 2000, and 2) the steady expansion in area irrigated by the Sardar Sarovar Project (SSP) canals, which have started supplying water to the water-scarce regions of north Gujarat. We will elaborate on this in the subsequent paragraphs.

Analysis of past growth trends for three important crops shows that with good monsoons, agriculture in Gujarat had grown substantially with steady expansion in either cropped area or yield growth. As against this, in drought years, the production has always suffered with shrinkage in area under irrigated winter crops, and sharp reduction in yield of kharif crops, including cotton and groundnut. In other words, 'criticality' of rainfall to sustain agricultural production in Gujarat is even higher as compared to the pre-Green Revolution period. The 4 consecutive years of good rainfall remarkably improved groundwater recharge, increased the storage in surface reservoirs throughout the state and improved soil moisture conditions. The reduced pressure on aquifers for irrigation due to availability of water from surface reservoirs reduced irrigation water requirement for crops due to improved soil moisture regime, and increase in replenishment together made a huge positive impact on groundwater balance, making more water available for subsequent years.

Second, the import of water from Sardar Sarovar reservoir through canals under SSP had in the very recent years boosted the agricultural production at least in a few districts of south (Bharuch, Baroda and Narmada districts) and north Gujarat (Ahmedabad and Gandhinagar). Though the distribution and delivery canals are still not ready for delivery of water to the fields in the entire command (of 1.8 m. ha), the length of the completed network is reasonable enough for farmers in many areas to tap water from the system. Table 2.3 provides the overall progress in the construction of canal network of SSP.

Table 2.3 Overall Progress in the Construction of Canal Network of SSP

Type of Canal	*Total Length When Completed (km)*	*Length** (km)*	*Physical Progress (%)*
Main Canal	448	448	100.00
Branch Canals	2,585	1,773	68.60
Distributaries	5,112	1,497	29.20
Minors	18,413	4,941	26.80
Subminors	48,058	10,055	20.90
Total	74,626	18,724	25.00

Source: Desai and Joshi, 2008, ** As of March, 2008

The total volume of water utilized from Narmada Canal System in the initial phase of the command as of March 2008 was 1,800 MCM. The gross area irrigated by this could be in the range of 2.4–3.27 lac ha, depending on an assumed delta of maximum 30 inches (750 mm) to a minimum of 22 inches (550 mm). In addition to this, Narmada Main Canal discharges water into several rivers of central and north Gujarat en route to Rajasthan. They are Heran, Orsang, Sherdi, Dhadhar, Saidak, Watrak, Meshvo, Sabarmati, Khari, Rupen, Banas, Pushpawati and Saraswati (Desai and Joshi, 2008). To exploit the situation, farmers put up engines to lift water from the canals and rivers and transport it to the fields. The area under wheat and cotton in the area around Narmada Main Canal has dramatically gone up during the past few years. The bumper production in cotton and wheat achieved in the recent years is a testimony to this. Narmada waters have also started producing several indirect benefits by replenishing the aquifers and raising the water table, as the rivers that are receiving Narmada water are in the alluvial basin with dewatered aquifers.

What a project like Sardar Sarovar can do to the semi-arid, alluvial areas of the state that are expected to receive water from its canal network for irrigation, can be guessed from a quick assessment of the agricultural scenario in south Gujarat. It is agriculturally one of the most prosperous regions of India, and is also socio-economically forward. What characterizes the region's agriculture are the two water-abundant gravity irrigation schemes, viz., Ukai-Kakrapar and Mahi. With the introduction of canal water, the farmers of the region have taken up cultivation of paddy-wheat, and cotton-wheat and perennial cash crops such as sugarcane and banana. The irrigation from canals has augmented the groundwater. The farmers who do not receive canal water are able to dig shallow tube wells and use groundwater for irrigation. The two schemes together irrigate around 5.20 lac ha of land. Hundreds of thousands of farmers in the area purely depend on canal water for irrigation. The paddy, sugarcane and cotton yields are one of the highest in the region. The continuous replenishment of groundwater enables farmers' easy access to well water for irrigation, as the water table is very shallow. In years of reduced inflows into the reservoirs, or the area not receiving sufficient rains, the farmers could still grow the traditional and high-valued crops using the groundwater, which is available in plenty.

When canal water is supplied to an area facing groundwater overdraft, apart from augmenting surface irrigation, it can reduce the stress on aquifers and do 'groundwater banking' for bad years. Hence, attempts to estimate the water productivity in canal-irrigated crops by merely looking at the total volume of water supplied by canals and the economic value of outputs generated from canal-irrigated fields would be highly misleading. The two additional benefits are the economic value of the outputs that can be generated from the replenished groundwater, and the positive externalities it induces on the cost of groundwater abstraction and the environment by raising the water table (Shah and Kumar, 2008).

Future of agricultural growth in ecologically fragile regions: Lessons from Gujarat

In spite of the constraints induced by poor water endowment, Gujarat had achieved significant strides in agriculture through modernization, diversification, good infrastructure for production and marketing. This is particularly significant for milk, horticultural crops and cash crops. But the spurt in the recent years (after 1998–99) has become very erratic and vulnerable to droughts, with sharp falls in production in such years. The phenomenon throws up some very important lessons for other fragile agro-ecologies of India. The state's water and energy policy had catalyzed uncontrolled exploitation of groundwater with mining in many areas. The state hadn't put any serious thinking on the sustainable use of its water resource so far. In the process it had used up all its renewable water resources, both surface and underground, and also most of the groundwater stock available in the alluvial basins of semi-arid areas. The 'criticality' of rains to the state's agriculture had become greater than ever before.[3] This is a dangerous situation.

The state's agricultural policymakers, for quite some time, believed that one way to protect the economic interest of the farmers is to subsidize electricity and provide good quality power and subsidized canal water. The policy of providing subsidized electricity to the farm sector was a wrong one.[4] Instead, there is a need to introduce metering and charging for every unit of electricity consumed (IRMA/UNICEF, 2001; Kumar, 2005). Only this can motivate farmers to use water and electricity efficiently. Findings of the research, discussed in Chapter 5, show that when confronted with marginal cost of using electricity, farmers tend to use electricity and water more efficiently, by improving physical efficiency of water use, by allocating water to crops that give higher returns from every unit of water and by improving the entire farming system, resulting in overall farming system water productivity. The recent decision of the state government to make issuance of new power connections for agro wells contingent upon farming agreeing to install meters and pay on pro rata basis is a welcome step to undo the damage.

As a result of this new policy, around 45% of the agricultural connections in the state are already under pro rata tariff.

There is no doubt that the initiative of the state government to promote micro irrigation systems, through its agency called Gujarat Green Revolution Company, had paid good dividends. The area under MI systems had increased to around 4.07 lac ha in 2010. Of this, 2.26 lac ha were under drips alone (Sankaranarayanan et al., 2011). The total area added to MI in the state during the period from 2001–02 to 2004–05 was less than 9,000 ha (Kumar et al., 2008c). Electricity tariff reforms will change the situation altogether for the better, with much greater adoption of drips and sprinklers, as farmers would be concerned with the use of every drop of water they pump out from underground. While certain policies had fueled agricultural growth in the entire state for more than a decade, it now appears that such a growth would be unsustainable. Agriculture has become highly vulnerable to the occurrence of meteorological droughts. The state now has taken major step of transferring water in bulk from the water-rich, land-scarce regions of south Gujarat to water-scarce, land-rich regions of north Gujarat and Saurashtra to reduce this vulnerability.

Such interventions to reduce drought vulnerability through improving water security positively impact human development and economic growth. The strong correlation between annual GDP growth rates and rainfall in East Africa observed over a period of 3 decades illustrates the criticality of irrigation in boosting rural economy and growth in semi-arid and arid tropics. As shown by Kumar et al (2008b), improving water security of a region, expressed in terms of sustainable water use index (SWI), improves the human development index (HDI) by reducing mortality and malnutrition.

While groundwater was a 'drought buffer' in agriculturally prosperous semi-arid and arid regions, the depletion of the very resource due to its unsustainable use is now posing a threat to future agricultural growth. This is now also evident in many parts of Rajasthan, Punjab, Andhra Pradesh, Tamil Nadu, Karnataka and Madhya Pradesh. The yield of wells is declining sharply in hard rock areas, with an increase in the number of wells not adding to well-irrigated area (Kumar, 2007). In Saurashtra, for instance, the well-irrigated area had declined in many districts. The situation is even more critical in areas that do not receive surface water resources.

To overcome its groundwater crisis, Gujarat government launched a massive program of decentralized groundwater recharge. This seems to have been driven by the general notion that more structures meant more water. There were no hydrological and economic considerations involved in planning water-harvesting systems in any of the basins. Most of these interventions were concentrated in basins that are already 'closed,' leading to dividing of the water rather than its augmentation. Often, structures are oversized (Kumar et al., 2008a). Also, evaporation losses from impounded water increased due to increase in reservoir area (Kumar et al., 2006). All these lead to poor economics. The states like Gujarat that are facing groundwater depletion problems should now look beyond such piecemeal solutions, and try to tackle groundwater depletion through long-term, institutional and policy measures.

While water harvesting, large water systems and water imports all have a place in water management, the chances of achieving desired results from the same would depend on the basin-wide hydrological planning. The catchment and basin hydrology needs to be studied and the scale, at which various small water harvesting systems and large water systems can be taken up in the basin, needs to be assessed based on proper estimates of dependable yield of the basin and the water demands. Potential for water demand management in agriculture, with diversification of cropping system to accommodate water efficient crops, and use of micro irrigation systems also needs to be explored. Further deficits can be filled through water imports, like what has been done in the case of Gujarat under the SSP. Only such an approach can ensure sustainable and equitable use of basin water resources for sustaining agricultural growth.

Notes

1. Here again, we did not consider the year, 2007–2008 (the year considered by Shah et al., 2009) for our estimates, as that year corresponds to a high growth year, with subsequent consistent decline in value of outputs in the 2 years that followed, i.e., 2008–2009 and 2009–2010.
2. The project is currently managed by Society for Integrated Land and Water Management (SOFILWM), Palanpur.
3. The reason is that the agriculture in the semi-arid and arid parts of the state is heavily dependent on normal monsoons not only as a source of critical moisture supply for kharif crops, but also as the source of recharge for the aquifers, as groundwater stocks are already exhausted. The other source of water for the people in these three regions is the storage in the minor, medium and large reservoirs.
4. The electricity pricing policy, which the state had followed for nearly 12 years, had contributed to the groundwater overexploitation, while triggering short-term growth and benefiting a few large well-owning farmers through electricity subsidies (Kumar and Singh, 2001; Kumar, 2007). The state's revenue loss through subsidy was to the tune of Rs. 4,100 crore in 2002–2003 alone (Government of India, 2002).

References

Bhalla, G. S. and Singh, G. (2009). Economic Liberalization and Indian Agriculture: A State-Wise Analysis. *Economic and Political Weekly. 44*(52), 34–44.

Bhatia, B. (1992). Parched Throats and Lush Green Fields: Political Economy of Groundwater in Gujarat. *Economic and Political Weekly.*

Desai, S.J. and Joshi, M. B. (2008). Narmada Water Plays Its Role-Capturing Initial Trends From Gujarat. Presentation, Sardar Sarovar Narmada Nigam Ltd., Gandhinagar, Gujarat.

Government of India (1996). *State-Wise and Crop-Wise Estimates of Value of Outputs From Agriculture (1980–81 to 1990–91).* New Delhi: Central Statistical Organization, Dept. of Statistics, Ministry of Planning & Programme Implementation, Government of India.

Government of India (2002, May). *Annual Report on the Working of State Electricity Boards and Electricity Department – (2001–02).* New Delhi: Planning Commission (Power and Energy Division), Government of India.

Government of India (2004). *State-Wise and Crop-Wise Estimates of Value of Outputs From Agriculture (1990–91 to 2001–02).* New Delhi: Central Statistical Organization, Dept. of Statistics, Ministry of Planning & Programme Implementation, Government of India.

Government of India (2008). *State-Wise and Crop-Wise Estimates of Value of Outputs From Agriculture (1999–2000 to 2005–06).* New Delhi: Central Statistical Organization, Dept. of Statistics, Ministry of Planning & Programme Implementation, Government of India.

Gulati, A. (2002). *Challenges to Punjab Agriculture in a Globalizing World.* Paper based on the presentation given at the policy dialogue on Challenges to Punjab Agriculture in a Globalizing World, jointly organized by IFPRI and ICRIER, New Delhi.

Gulati, A., Shah, T. and Shreedhar, G. (2009). *Agriculture Performance in Gujarat Since 2000: Can Gujarat Be a 'Divadandi' (Lighthouse) for Other States?* IWMI and IFPRI, New Delhi.

Ilyas, S. M. (1999). Water Harvesting Towards Water Demand Management and Sustainable Development. *Journal of Rural Reconstruction, 32*(2), 31–43.

Institute of Rural Management Anand/UNICEF (2001). *White Paper on Water in Gujarat.* Gandhinagar: Government of Gujarat.

Kumar, M. D. (2002). *Reconciling Water Use and Environment: Water Resources Management in Gujarat Resource, Problems, Issues, Options, Strategies and Framework For Action.* Report of the Hydrological Regime Subcomponent of the State Environmental Action Programme supported by the World Bank, prepared for Gujarat Ecology Commission, Vadodara.

Kumar, M. D. (2005). Impact of Electricity Prices and Volumetric Water Allocation on Energy and Groundwater Demand Management: Analysis From Western India. *Energy Policy, 33*(1), 39–51.

Kumar, M. D. (2007). *Groundwater Management in India: Physical, Institutional and Policy Alternatives.* New Delhi: Sage Publications.

Kumar, M. D. and Amarasinghe, U. (Eds.). (2009). *Water Productivity Improvements in Indian Agriculture: Potentials, Constraints and Prospects.* Strategic Analysis of the National River Linking Project (NRLP) Series 4, Colombo: International Water Management Institute.

Kumar, M. D., Ghosh, S., Patel, A. and Ravindranath, R. (2006). Rainwater Harvesting and Groundwater Recharge in India: Critical Issues for Basin Planning and Research. *Land Use and Water Resources Research, 6*(1), 1–17.

Kumar, M. D., Patel, A. R., Ravindranath, R. and Singh, O. P. (2008a). Chasing a Mirage: Water Harvesting and Artificial Recharge in Naturally Water Scarce Regions. *Economic and Political Weekly, 43*(35), 61–71.

Kumar, M. D., Shah, Z., Mukherjee, S. and Mudgerikar, A. C. (2008b). *Water, Human Development and Economic Growth: Some International Perspectives.* Proceedings of the 7th Annual Partners' Meet of IWMI-Tata Water Policy Research Program, International Water Management Institute, South Asia Regional Office, ICRISAT Campus, Hyderabad, April 2–4.

Kumar, M. D. and Singh, O. P. (2001). Market Instruments for Demand Management in the Face of Scarcity and Overuse of Water in Gujarat, Western India. *Water Policy, 5*(3), 387–403.

Kumar, M. D., Turral, H., Sharma, B., Amarasinghe, U. and Singh, O. P. (2008c). Water Saving and Yield Enhancing Micro-Irrigation Technologies in India: When and Where Can They Become Best Bet Technologies? In M. D. Kumar (Ed.), *Managing Water in the Face of Growing Scarcity, Inequity and Declining Returns: Exploring Fresh Approaches: Vol. 1* (pp. 1–36). Proceedings of the 7th Annual Partners' Meet, IWMI-Tata Water Policy Research Program, April 2-4, IWMI South Asia Sub-Regional Office, ICRISAT Campus, Patancheru, Andhra Pradesh.

Kumar, M. D. and van Dam, J. C. (2013). Drivers of Change in Agricultural Water Productivity and Its Improvement at Basin Scale in Developing Economies. *Water International, 38*(3), 312–325.

Pearce, D. W. and Warford, J. (1993). *World Without End: Economics, Environment and Sustainable Development.* New York: Oxford University Press.

Planning Commission (2008). *Eleventh Five Year Plan, 2007–2012, Agriculture, Rural Development, Industry, Services and Infrastructure.* New Delhi: Government of India.

Ranade, R. and Kumar, M.D. (2004). Narmada Water for Groundwater Recharge in North Gujarat. *Economic and Political Weekly, 39*(31), 3510–3513.

Ray, S. and Bijarnia, M. (2006, July 10). Upstream vs. Downstream: Groundwater Management and Rainwater Harvesting. *Economic and Political Weekly, 41*(23) 2375–2383.

Sankaranarayanan, K., Nalayani, P., Sabesh, M., Usharani, S., Nachane, R. P. and Gopalakrishnan, N. (2011). Low Cost Drip: Low Cost and Precision Irrigation Tool in Bt Cotton. Technical Bulletin No. 1/2011, Central Research Institute for Cotton, Coimbatore Station, Coimbatore.

Shah, T., Gulati, A., Hemant P., Sreedhar, G. and Jain, R. C. (2009,). Secret of Gujarat's Agrarian Miracle After 2000. *Economic and Political Weekly, 44*(52), 45–55.

Shah, Z. and Kumar, M. D. (2008). In the Midst of the Large Dam Controversy: Objective, Criteria for Assessing Large Water Storages in the Developing World. *Water Resources Management, 22*, 1799–1824.

Thobani, M. (1997). Formal Water Markets: Why, When and How to Introduce Tradable Water Rights. *The World Bank Research Observer, 12*(2), 161–179.

3 Ghost workers and invisible dams[1]

Checking the validity of claims about impacts of NREGA

Nitin Bassi, M. Dinesh Kumar and A. Narayanamoorthy

Introduction

Rural poverty and unemployment in India have grown in an unprecedented manner during the last few decades. There is a growing incidence of rural youth shifting from agriculture into unproductive activities, compounding this problem. In order to reverse this trend and to provide livelihood security to the rural unemployed, the government of India (GOI) enacted the National Rural Employment Guarantee Act (NREGA)[2] in 2005. The act provides for 100 days of guaranteed employment to every rural household in a financial year for unskilled manual work. The Act, initially notified in 200 districts, at present covers 625 districts (nearly all the districts in the country) and claims to have benefited some 5 crore poorest households in the year 2011–12. With the budget allocation of Rs. 11,300 crore in 2006–07[3] under the umbrella of the NREGA, this is probably the largest rights-based social protection initiative in the world (Farrington et al., 2007). As per Schedule I of the Act, the work under National Rural Employment Guarantee Scheme (NREGS) will be essentially the creation of sustainable rural assets.

The NREGA builds on earlier experience with Employment Guarantee Scheme (EGS) in Maharashtra (Sjoblom and Farrington, 2008). The key component of NREGA is the provision of employment by the state at a prescribed wage for those unable to find alternative employment, which provides a form of social safety net to the rural unemployed people. The long-term objectives of the scheme include: a) enhancement of livelihood security in rural areas, b) creating productive assets, c) protecting the environment, d) empowering rural women and e) fostering social equity. Apart from affirming the 'right to work,' the Act also seeks to ensure that the poor have a say in decisions on the works to be undertaken, so that such works contribute to improvement in their livelihoods (McCord and Farrington, 2008).

There is, however, a dearth of systematic studies on the impact of employment guarantee schemes (EGS) and cash transfer (CT) schemes on the poor, especially in the South Asian region, which employed robust methodologies. Most such studies are available in Latin America and sub-Saharan Africa (Hagen-Zanker et al., 2011). Further, the impact of EGSs, which are based on creation of public assets, on regions' ecology and environment is hardly explored. Although, recently some scholars in India attempted to highlight the environmental and vulnerability reduction impacts of the NREGA (see, for instance, Tiwari et al., 2011), their methodology was based on several faulty assumptions, and analysis was marred by limited empirical data and poor understanding of the basic concepts of ecosystem services (Kumar et al., 2011). In lieu of this, this chapter attempts to: a) make quick assessment of the performance of implementation of water-management activities under NREGS; b) analyze the scheme design and investigate the employment-generation potential of the scheme; c) scrutinize the welfare effects of the water management activities chosen under the scheme and d) suggest the broader water-management strategies suitable for each one of the identified typologies.

Performance of water-management activities under NREGS

Works related to water and soil conservation, afforestation and land development were given top priority under the NREGS. The water-management (WM) works specifically includes a) water conservation and water harvesting; b) drought proofing; c) irrigation canals; d) provision of irrigation facility to land owned by households belonging to SC/ST or to land of the beneficiaries of land Reforms/ Indira *Awas Yojana*/BPL families; e) renovation of traditional water bodies; f) land development and g) flood-control and protection works (GOI, 2008). During the 3-year time period (2006–07 to 2008–09), more than 31.44 lac water-management-related works have been completed, with a total expenditure of 35.9 thousand crore (Table 3.1) (Sharma, 2009). Of these, the maximum number of works were undertaken on water conservation and water harvesting.

The statistics provided in Table 3.1 do not include the money, time and labor spent on uncompleted works. Further, the completion rate of various WM interventions (as a percentage of total works undertaken) does not show very encouraging results. Between 2008–09 and 2011–12, no significant improvement in work completion rate has been witnessed, with highest being achieved in land-development works (Figure 3.1).

Considering the nature and size of the WM works undertaken, many believe that NREGS would yield a remarkable impact on rural water management, providing water security in some water-deficit areas (see for instance Shah, 2009) and protecting some other areas from devastation caused by floods. However, initial evidence suggests that neither the nature nor the quality of assets created under the water-management works is satisfactory. The reason for this is quite clear. Little consideration is given to the social, economic and hydrological aspects while selecting different interventions. Focus of the WM works was more on creation of stereotype assets with little consideration for local relevance. In many instances, poor planning and lack of maintenance resulted in completed assets falling under disuse (World Bank, 2011; CSE, 2008). For example, water-harvesting structures were built without any provision for catchment protection (CSE, 2008). More importantly, in the name of storage enhancement, large amounts of earth are excavated from ponds and tanks in rectangular-shaped pits, without any attention being paid to the amount of silt in the tank, or the topography, leading to increased

Table 3.1 Water-Management Works Under NREGS (From 2006–07 to 2008–09)

Sr. No.	*Category of Water-Management Works*	*Type of Work Undertaken Under Each Category**	*Expenditure per Work Undertaken (000' Rs.)*	*Benefit Created per Work Undertaken*
1	Water conservation and water harvesting	Digging of new tanks/ponds, percolation tanks, small check dams	160.80	276.43 Cu Mt.
2	Drought proofing	Afforestation and tree plantation	147.06	3.68 Hectare
3	Irrigation canals	Minor irrigation canals	118.18	0.45 Km
4	Provision of irrigation facility to land owned by HH of SC/ST OR IAY/BPL beneficiaries	Digging of farm ponds	39.16	0.26 Hectare
5	Renovation of traditional water bodies	Desilting of tanks/ponds, desilting of old canals and traditional open well	207.10	804.73 Cu Mt.
6	Land development	Plantation and land leveling	73.44	3.12 Hectare

*List may not be inclusive

Source: Table compiled from data presented in Sharma, 2009

soil erosion in subsequent rains. All the attention is on increasing the amount of wage employment and making measurements easy, rather than regaining the original shape and storage capacity of the water body. Moreover, as a result of noninclusion of maintenance works under the ambit of the NREGS, ability of these structures to provide sustainable yields is under question.

From a poverty-reduction viewpoint, NGREGA offers too little as a public scheme. The types of WM activities for which work can be funded (e.g. water conservation, land development, afforestation, provision of irrigation systems, or flood control) are prone to being taken over by wealthier sections of society. Also, poor implementation of many of NREGS works has led beneficiaries to think that it is no better than any other government schemes that have had little impact on poverty (Sjoblom

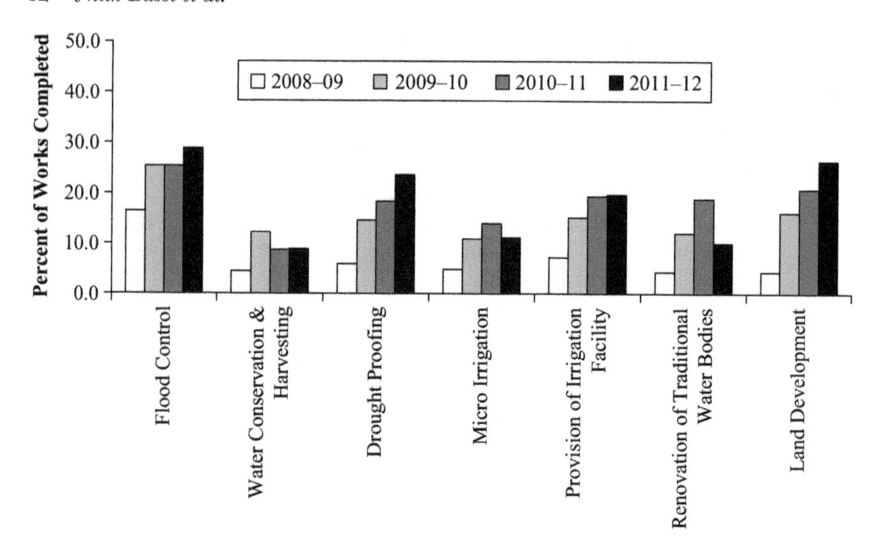

Figure 3.1 Year-Wise Completion Rate of Water-Management Works Under NREGS
Source: Data compiled from MoRD statistics and NCAER-PIF Study, 2009

and Farrington, 2008). Absence of proper public audits has further aggravated the problem. On the human resources development front, NREGS works don't provide skills enhancement and therefore do little to strengthen human capital. In addition, by taking work directly to the people, the scheme may discourage them from moving to more economically dynamic areas (Farrington et al., 2007).

Why employment generation under NREGS is poor

It has been reported from many states in India such as Punjab, Kerala and Maharashtra that performance of NREGS in terms of employment generation is very poor. Singh and Gill (2010) reported that in Punjab, only 28% (in 2008–09) and 38% (in 2009–10) of the job card holders were provided with employment. Further, more than one half of jobs created were confined to only five districts of the state. But, there is no systematic analysis of the root cause of this problem of abysmal performance of the scheme. Instead of analyzing the problem, the government is keen to increase the allocation for NREGA to create a greater impact on poverty. The key argument by the policymakers and the scholars alike (see Shah, 2009, for instance) is that NREGS benefits only the poor and the work-linked payment reduces the error of inclusion of the undeserving in the scheme. Perhaps this is merely based on the observation that the number of households reporting for work is much smaller than that of job card holders. But, for this argument to hold water, three conditions need to be satisfied.

First: There should be sufficient employment opportunities in the villages for all job card holders that are created through public works. *Second*: There should be sufficient demand for the low-paid wage labor under NREGA from the really deserving ones, or in other words NREGA work should offer sufficient incentive for the poor people, who are dependent on year-round wage labor for their survival. *Third*: Employment from NREGA should be available during the lean season for the wage laborers.

Unfortunately, none of these conditions are met in NREGA program. The employment potential of villages in terms of number of days of wage labor that the villages can create through public works such as construction/renovation of water bodies, road construction, etc., has never been studied. Nor has there been any real assessment of the unemployment situation in the rural areas across the country. First, availability of land for undertaking public works is open to question, particularly in more agriculturally prosperous states. Out of India's total geographical area (which is about 329 m. ha), in 2007–08, only 13%, i.e., 42.4 m. ha, was barren and fallow. Against this, the agricultural prosperous state of Punjab has only 1% (Tiwana et. al., 2007), and Kerala has about 3.6% of total geographical area that is barren and fallow land (GOI, 2013). The situation is even more precarious in Bihar, which also has a high rate of rural poverty and unemployment.

Thus, whether deserving or undeserving, people are left out of getting any employment, as there is too little of land available for taking up any public works such as pond digging or construction of roads. Further, activities such as desilting of ponds, cleaning of village drains and minor canals, etc., undertaken in various states would not result in 100 days of employment for each of the registered job card holders (Figure 3.2). At a national level, only 8% of households provided with work completed 100 days of employment during 2011–12. But villagers are still happy to get job cards, as after 15 days of application for work, they become eligible for a daily unemployment allowance (GOI, 2008), in case no employment is

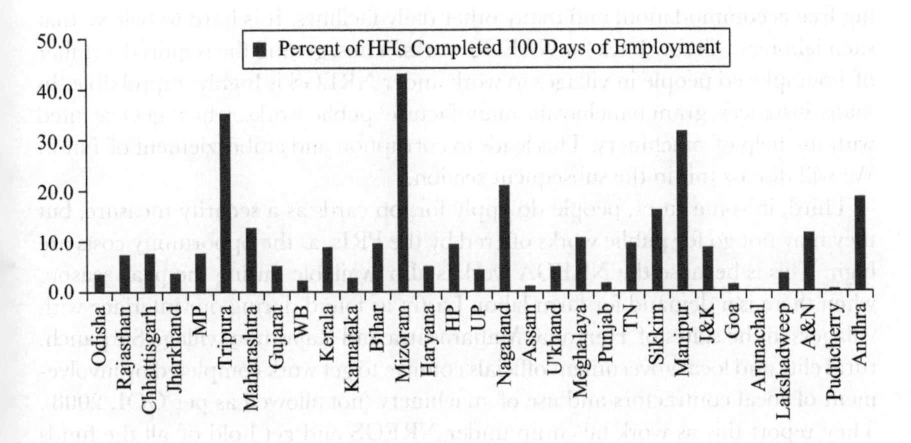

Figure 3.2 State-Wise Proportion of Households Completing 100 Days of Employment

Source: Authors' own analysis using MoRD statistics

offered to them. There is no reason to believe that only the most deserving among the job card holders demand jobs or at least demand compensation for unemployment. It is quite likely that the poor, unemployed people, instead of waiting for opportunities in their village, either go for work in other farms or migrate to towns and cities in search of regular labor.

Second, it is hard to find sufficient unemployed people in the villages who are willing to work for NREGS and avail of the benefits. According to some estimates, there are about 3.2 crore people in rural India who are unemployed and looking for jobs (GOI, 2010). In the financial year 2010–11, funds available under NREGS for paying wages were around 32,000 crore (also includes carried-over balance of previous year). Considering that each employed unskilled laborer gets Rs. 10,000 in a year, some 3.2 crore jobs were required to be created in 2010–11. Thus, even if these official statistics of unemployment rate are reliable, NREGS still has sufficient funds to provide 100 days of work to all the unemployed people in the villages, provided the allocation across states reflects the employment situation in these states.

But in reality, this is not so. In many cases, fund allocation is far in excess of the genuine needs of the states. Even in states like Bihar where unemployment is high, large numbers migrate to other states, which are economically prosperous and where labor is in short supply, to work on farms and elsewhere instead of seeking wage labor under NREGA.[4] This dichotomy is because what NREGS offers in terms of income gain for laborers is too little to make respectable living. Here it is important to note that the employment generation through NREGA is generally poor in densely populated states like Bihar, which have extremely low per capita land availability. As per one estimate, annually about 52 lac poor people from Bihar alone migrate to other states in search of better livelihood opportunities. NREGS has actually adversely affected the labor market by providing migrant laborers with better and sometimes unfair bargaining power. As was found in the state of Punjab, in addition to increased wages, agricultural laborers were demanding free accommodation, and many other daily facilities. It is hard to believe that such laborers will stay back for NREGS works. Thus, getting the required number of unemployed people in villages to work under NREGS is highly improbable. In many instances, gram panchayats 'manufacture' public works, which get executed with the help of machinery. This leads to corruption and embezzlement of funds. We will discuss this in the subsequent section.

Third, in some cases, people do apply for job cards as a security measure, but they may not go for public works offered by the PRIs, as the opportunity costs are high. This is because the NREGA work is also available during the peak season, when there is a demand for farm labor. Lastly, as found during our interface with villagers in the states of Haryana, Maharashtra and Rajasthan, village Sarpanch, rural elite and local government officials connive to get work completed by involvement of local contractors and use of machinery (not allowed as per GOI, 2008). They report this as work taken up under NREGS and get hold of all the funds that are actually allocated to the poor and the needy. In return, they offer only a small proportion (10–20%) of the money to job card holders. Such practices are quite common, as at least 50% of the works has to be allotted to gram panchayat for execution, as per NREGA. Singh and Gill (2010) and Kumari (2010) identified some other practical problems, which have resulted in poor employment generation under NREGS. These problems include lack of transparency; absence of social audit, proper project planning and professional staff; delayed payments; fake muster rolls and fund locking.

Thus, faulty design of NREGS, which does not consider the employment-generation potential in the villages through public works, the number of unemployed people in the villages and their demands vis-à-vis wage rates and the labor dynamics, leads to over-allocation of funds, and the resultant manufacturing of works by the panchayats, embezzlement of funds and the ineligible claiming the benefits of the scheme, while the most deserving lose the opportunity. Obviously, the solution does not lie in allocating more days of labor, but in targeting NREGS funds to areas where the rural employment scenario is poor, and where avenues exist for generating employment through public works.

Emerging concerns

Implementing water-management works under NREGA, on the scale envisaged, has posed major challenges. Corruption in the implementation is rampant. Till March 2010, some 1,138 complaints related to irregularities in implementation of NREGA activities were received, with the maximum reported cases from the state of Uttar Pradesh (Figure 3.3). Field evidence suggests that spending on some of the water-management works has not only been inadequate but also unwise. For instance, ponds have been dug in areas with scant rainfall, without conceptualization of factors such as catchments areas and sources of recharging (NCAER-PIF, 2009). As a matter of fact, residual catchments are hard to find in naturally water-scarce regions, where already a large number of small and large water-impounding structures exist, including those that are both traditional and modern. The flows generated from the natural catchments are already committed for the small and large reservoirs downstream (Kumar et al., 2008). As a result, construction of new structures in the upper catchments produces negative effects downstream, in the form of reduced flows into tanks and reservoirs (Bachelor et al., 2002; Kumar et al., 2008). It is important to mention here that none of the small river basins in the country are gauged (Kumar et al., 2006), with the result that there are no official data on their runoff generation potential. Decentralized water-harvesting efforts at the village level, which are uncoordinated at the level of the large watersheds or the river basin, therefore result in overappropriation of surface water from catchments.

No geo-hydrological investigations are undertaken for initiating activities that are intended to be groundwater recharge schemes. Not only is the provision of funds for doing such investigations an issue, but also the availability of scientific and technical manpower on such a large scale in rural areas is questionable. These issues are well captured in the latest World Bank study on 14 social protection programs in India, carried out for the Planning Commission of India,

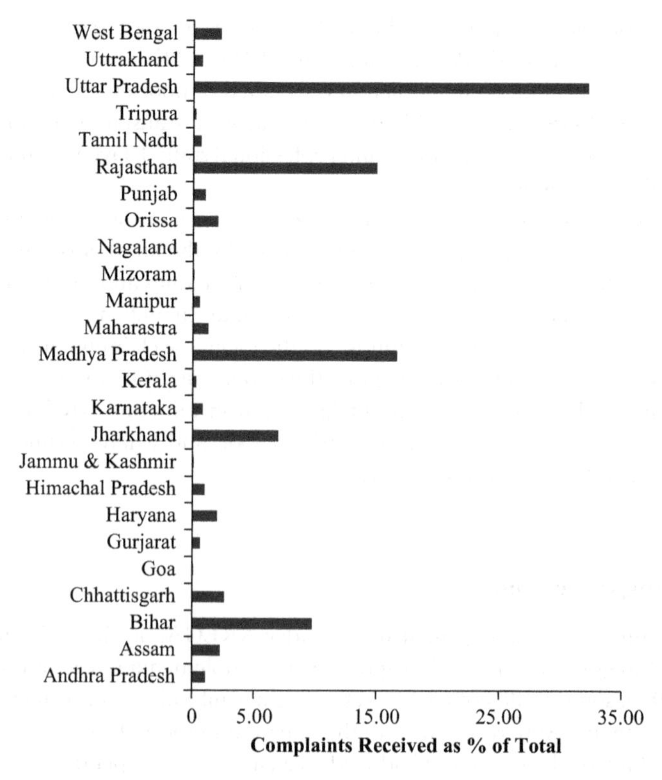

Figure 3.3 Status of Complaints (Related to Irregularities in Implementation of NREGA Activities) Till March 2010

Source: Data compiled from NREGA Complaints Sheet, MoRD, GOI

which made serious remarks about the poor quality of work carried out under NREGA, relating to water harvesting and conservation. As World Bank (2011) notes:

> A key constraint to building high quality assets is the lack of technical support to communities as input to planning MGNREG works (e.g., through resource mapping exercises) as well as the shortage of technical staff in designing and supervising works. A large number of works, particularly those related to water conservation, remain incomplete, either due to lack of technical support to GPs or the onset of monsoons. (World Bank, 2011, p. 84)

In some cases expenditure was found to have been incurred on nonexistent projects (NCAER-PIF, 2009). A recent such scam was unearthed by the Sarpanch of village Hitameta, Dantewada (Sharma, 2011). While one crore rupee was spent on four nonexistent stop dams, 122 workers out of the total 145 listed workers never worked at the site. The muster roll even brought back six deceased villagers to life by listing their names under the workers who have collected their wages. As per information provided by a senior IAS officer, nearly 40–50% of all government expenditure in Dantewada is lost in corruption (Sharma, 2011). This is just one of the myriad of frauds that are prevalent in NREGS projects.

At several places, emphasis is more on spending a larger amount of money than on ensuring quality works. According to Ambasta et al. (2008), in some districts of Chhattisgarh, planting was done but no provision was made for either watering or protection (mainly from grazing) of plantation, whereas in few districts of Madhya Pradesh, farm bunding had been initiated without any proper technical planning. In some other tribal districts of Chhattisgarh, works were focused mainly on activities for which standardized estimates were available. Thus, plans are made and approved for implementation in a 'top down' manner. As a consequence, major portions of the approved funds were utilized on roads, where drought proofing should have been given top priority considering the area to be one of the poorest tribal pockets of the country, with a long history of droughts. The hydrology and topography of these areas is naturally suited to watershed works too, but these remained far off from the priority of NREGA plans in the state (Ambasta et al., 2008). Such poor implementation strategies stem from lack of understanding of the interaction between various components of the hydrological system, viz., surface water, groundwater and catchments; and various environmental resources such as land, water and trees.

Further, failure to understand the rural labor markets has led to serious negative impacts on food security for the small and marginal farmers. It has been argued that NREGS led to withdrawal of sections of the labor force to work on projects of uncertain value that resulted in market distortion (wage increase). Agricultural activity of those who rely on hired workers, including small and marginal farmers who survive on small margins of profits, has also been affected (Panagariya, 2009). The plight of Punjab farmers, who are heavily dependent on farmhands from eastern and central India to work on their fields, has been reported. Since they have stopped getting laborers from central India, the demand for laborers from eastern India has gone up. This shortage of migrant laborers from central India is attributed mainly to the jobs created back home under the NREGS. As a result of the increased demand for labor, the seasonal wage rate in Punjab has increased threefold from a mere Rs. 700 to Rs. 2,000–2,500 per acre, in just about 2 years (Sharma, 2010). As per the latest data available from the Labour Bureau, between December 2009 and December 2010, farm wages have increased by around 43% in Orissa. In states such as Uttar Pradesh, Karnataka and Tamil Nadu, daily wages for plowing, sowing, weeding, transplanting and harvesting rose by nearly 25% (Gupta and Sidhartha, 2011).

Effectiveness of water-management interventions

Integrating hydrological and economic consideration in planning water-related works is extremely crucial for the success of NREGS in terms of reducing negative welfare effects and improving land and water management by augmenting water resources in rural areas that are capable of strengthening food and livelihood security (Bassi and Kumar, 2010).

With increasing natural and man-made disasters, flood protection and control have become increasingly important. India has been tackling the problem of floods through structural and nonstructural measures. While nonstructural measures like flood forecasting aim at improving the preparedness for floods, structural measures involve the construction of embankments, dams, drainage channels and reservoirs that prevent floodwaters from reaching potential damage centers (Gupta et al., 2003). However, these efforts have provided little solace and there is a recurrent large-scale economic and human loss during floods in the Gangetic and Brahmaputra basins. Based on the analysis of data for three highly vulnerable states in the eastern region of India, Gupta et al. (2003) argued that flood protection measures have been inadequate in controlling losses and reducing vulnerability.

Understanding flood management requires study of hydrology, open channel hydraulics, and river morphology. Constructing embankments for flood control is one of the cheapest and fastest executable options for flood protection. However, embankments can create drainage difficulties in the countryside and induce a sense of security that reduces the level of alertness amongst the populace (Pandit, 2009). Thus, embankments require careful maintenance works and stabilization works. Stabilization works can be done through plantation of trees on the embankments. Dams and reservoirs are the best flood control structural measures, but they can only be constructed by a skilled workforce under the supervision of trained professionals. Therefore, they cannot be within the purview of NREGA, as it stands today. However, measures to stop soil erosion and silting up of existing reservoirs can be undertaken as a part of scheme WM works.

As regards water-harvesting interventions, there is a growing concern over their economic viability and downstream impacts. Planning of local water-harvesting/groundwater recharge schemes is not backed by proper hydrological and economic analysis. Therefore, there is hardly any knowledge of the cost per cubic meter of the harvested water through such schemes. This issue of economic viability becomes far more serious as these schemes are largely implemented in semi-arid and arid regions. The reason is that these regions have extremely limited runoff potential, with high interannual variability. Studies show that rainwater harvesting has limited potential to reduce the demand–supply imbalances and provide reliable supplies (see for instance Kumar et al., 2006; Kumar et al., 2008), and is economically unviable (Kumar et al., 2008) in water-scarce regions. Further, in these water-scarce regions, runoff harvesting does not offer any potential for augmenting the supplies at basin level, and instead creates huge negative economic (Kumar et al., 2008; Ray and Bijarnia, 2006), social and environmental externalities (Kumar et al., 2008) downstream. Harvested rainfall may also increase depletion, reducing downstream users' access to water from streams or groundwater. This loss of downstream access to the water may be severe and irreversible if the entire water supply is consumed by crops or growth of other vegetation (Baron, 2009).

Nevertheless, there were recent efforts by scholars to romanticize NREGA interventions from rural water and environmental management points of view (see Shah, 2009; Tiwari et al., 2011). For instance, Tiwari et al. (2011) highlighted role of water harvesting and conservation work under NREGS in environmental service enhancement and vulnerability reduction. However, the entire analysis was based on extremely limited and pseudo-empirical data, faulty assumptions and fictitious methodologies.[5] That said, water harvesting at the local level would make better sense if scientific inputs were considered in the planning of these interventions. These inputs include: analysis of rainfall intensity and pattern; reliable estimates of runoff from the catchments; analysis of engineering properties of the soils; topography; and, geo-hydrological data including geo-hydrological parameters of the formations, mapping of geological structures and groundwater–surface water interactions (Kumar et al., 2006). This would help in designing optimally sized structures, thereby saving the scarce financial resources.

Further, at the basin level, such schemes would need careful hydrological and economic planning, including assessment of unutilized runoff available for further harnessing. This should lead to determining the optimum number of structures that can be built to have positive hydrological and economic impacts. Unplanned and unscrupulous building of water-harvesting structures by village communities without due consideration to the basin's uncommitted stream flows would only result in redistribution of water across the basin, with negative social and ecological consequences for downstream areas, as they dry up local streams and reduce the flows into existing reservoirs meant for irrigation and drinking purposes. The recent evidence from Saurashtra (Aji reservoir in Rajkot) and Rajasthan of drying up of public reservoirs is just an indication of the larger menace if necessary caution is not exercised in implementing water-harvesting schemes in villages. In fact, in 2006, the government of Rajasthan threatened legal action against the government of Madhya Pradesh for the latter's refusal to dismantle hundreds of illegally constructed check dams/water-harvesting structures (WHS) along the Chambal River in the Malwa region. The Rajasthan government felt that due to these structures, the inflow of water into the Gandhi Sagar Dam has reduced. As per one estimate, the total number of WHS constructed in the catchment area exceeds 1,500 and includes structures that were built by villagers and NGOs under watershed projects (*The Indian Express*, 2006). Contrary to the concerns of downstream users, under NREGS, several hundreds of villages from the same region embark on activities like the digging up of tanks and ponds in the villages, without recognizing the aggregate impacts on downstream water bodies.

There is a general belief that afforestation would improve hydrological regimes in water-scarce regions. This is a dangerous misconception. Trees require water for physiological processes in the form of evapo-transpiration (ET) more than conventional field crops, and saplings require artificial application of water for their protection in semi-arid and arid areas, which experience erratic rainfall, as their limited root system does not allow them to take water from deeper soil strata (vadoze zone). Soil moisture for survival of saplings is difficult to manage in semi-arid areas on a sustainable basis. There are two reasons for this: a) Afforestation activity is generally taken up in wasteland and pasture land, which are away from prime agricultural land, making the access to irrigation sources difficult; and b) Most semi-arid regions in India experience very high variability in rainfall and rainy days, and generally years of low rainfall are characterized by fewer rainy days, which means long dry spells in years of low rainfall (Kumar et al., 2008; Pisharoty, 1990). Hence, there are two predictable outcomes of afforestation programs in such cases: First, overall survival rate of saplings becomes very low. Second, we deplete the available freshwater supplies to keep the survival of saplings high.

On-farm water management and renovation of traditional water bodies can be suitable for drought proofing in these semi-arid and arid regions. Normally, any meteorological drought will translate into a hydrological drought marked by reduced surface water and resultant negative impact on groundwater recharge (Kumar et al., 2009). Therefore, the available water will have to be used more efficiently. On-farm water management will prove to be beneficial for drought proofing. Under NREGS, technical interventions like field leveling and rehabilitation of earthen canals including their desilting and lining will be more suitable for reducing wastage of irrigation water, thereby achieving water-demand management. It is important to mention here that in arid and semi-arid regions, prevention of water wastage occurring through deep percolation of irrigation water in the field can potentially lead to real or 'wet' water saving (Kumar, 2009). So is the case with prevention of seepage from canals. (Please see Seckler, 1996, for definitions of 'dry' and 'wet' water saving.)

Normally, under the name of renovation, capacity enhancement work of traditional water bodies such as tanks and ponds, including digging of earth and raising embankments, is taken up. Mostly, earth-moving machinery is employed in executing this work, while, the earthwork like this actually gets reported under human labor. Our recent fieldwork in the Pali district of northern Rajasthan suggests that there is widespread manipulation of the work at sites under NREGS. The nexus between the laborers enrolled for NREGA work and the local contractors make it possible for laborers to claim wages on the basis of the volume of earthwork completed on the ground, which is actually executed with the use of large earth-moving machinery. The laborers in turn pay a portion of the receipts to the contractors based on the machine labor employed. In reality, the laborers never go to the site and execute the work themselves.

This practice of using moving machinery has two negative outcomes: 1) There is overdoing of earthwork. Unlike with human labor, scrapping of silt is not possible with earth-moving machinery. 2) The original shape of the water body is permanently lost, and the embankment slope is destabilized, resulting in increased rate of siltation of the water body. On the other hand, such activities do not help in improving water management during droughts, as tank inflows will be much less during drought years as compared to normal years. Instead, what is required is embankment protection using compaction and pitching and construction of waste weirs. Breaching of tank embankments and flooding of surrounding fields is a common scene in south Indian villages. Serious irregularities in the implementation of NREGS in Rajasthan and massive corruption that threatens to destroy the real aims of rural decentralization has also been acknowledged by Aruna Roy, member of National Advisory Council (*The Times of India*, 2010).

Strategies for effective implementation of water-management activities under NREGS

Bassi and Kumar (2010) identified three broad typologies to determine the priorities vis-à-vis the type of land- and water-based interventions that take into agro-climatic, hydrological and geological factors and that are appropriate for NREGA (refer to Table 3.2). Based on these typologies and discussion on the effectiveness of WM interventions in the earlier section, the following strategies have been proposed.

Table 3.2 Broad Typologies Based on Water Situation of the Region

	Type	*Region Included*	*Climate*	*Soil Type*	*Hydro-geology*
Typology I	Naturally and physically water-abundant regions	Gangetic Plains region of Bihar, Uttar Pradesh and West Bengal; Eastern Plateau region; East Coast Plains region of Orissa	Sub-humid climate with heavy rainfall ranging from 1,000 to 2,000 mm	Mainly alluvial and in some places red and yellow deltaic	Unconsolidated formations with rainfall infiltration factor ranging from 0.08 to 0.25
Typology II	Naturally water-abundant but physically water-scarce regions	Western and Eastern Himalayan region; West Coast Plains and hilly region of Maharashtra, Karnataka and Kerala	Humid climate with very heavy annual rainfall, some areas receiving more than 5,000 mm	Mainly hill alluvial and brown hill soil	Mostly hilly areas with steep slopes and semi-consolidated formations characterized with low ground water potential
Typology III	Naturally and physically water-scarce regions	Punjab, Haryana and Rajasthan; Central, Western and Southern Plateau region; Gujarat Plains and Western Dry regions	Arid and semi-arid climate with very low average mean annual rainfall (400–600 mm), even 100 mm at few places	Black soil, red sandy, coastal and sandy alluvial	Characterized by presence of consolidated and semi-consolidated formations. Rainfall infiltration factor is as low as 0.01 in some areas.

Source: Adapted from Bassi and Kumar, 2010

Flood control and protection works as WM intervention will be highly effective in regions that are both 'naturally water abundant' and have surplus water resources from the point of view of availability and demand. These areas are characterized by high to moderate rainfall, easy access to the available water resources (both surface and groundwater) and low water demands. However, adequate financial resources to access water are not available with the populations living there, causing economic water scarcity. Most parts of eastern India fall under this category. The flood control and protection works that can be undertaken include embankment construction, embankment stabilization through afforestation, and measures to stop soil erosion and silting up of existing reservoirs (Bassi and Kumar, 2010). A recent study from Bihar shows that social afforestation program undertaken by the state forest department is extremely successful in six participating districts not only from the point of view of developing wastelands but also from the point of view of providing gainful employment to the rural landless families (see Gupta, 2009). The villagers can be given the rights to harvest timber from the forest, once the trees mature. From a hydrological point of view, creating a forest cover improves the hydrological regime, by reducing the peak runoff rates and increasing the evaporation from the shallow groundwater. This can have some positive effects on controlling floods in regions like the Gangetic-Brahmaputra basins, and reducing problems of water logging in high water table areas.

Works related to runoff water harvesting and provision of irrigation facility to land owned by economically weak classes can be best suited for regions that are naturally water rich but physically water scarce. This region covers northern and northeastern India. The landscape in these areas is mostly hilly and even after receiving good rainfall, much of the water is lost as runoff. Such regions are perfect for surface water harvesting or creating impoundments. But, before undertaking these interventions, proper geo-technical studies should be carried out to prevent undesirable consequences such as landslides. Soil and water-management interventions including continuous bunding, drainage control and gully plugging (watershed-management approach) are equally effective in these regions. Small on-farm storages can also be constructed in these hilly areas to provide irrigation facilities on the farmers' agricultural field.

Renovation of traditional water bodies and on-farm water management can be an important intervention in naturally water-scarce areas, which experience physical water scarcity, owing to water demands far exceeding the total renewable water resources. Such regions in India are characterized by variable rainfall and high evaporation rates. Most parts of western, northwestern, central and peninsular India fall under this category (Kumar et al., 2006; Kumar et al., 2008). The agricultural water demands are very high in these regions, with large irrigated areas and high evapo-transpiration (Kumar et al., 2008). Since management of agricultural water demand is extremely important for mitigating water scarcity in these regions, investments should be on land leveling for on-farm water management, desilting and lining of canals, and renovation of traditional water bodies comprising silt scrapping, waste weir construction and embankment stabilization. It is also clear that such activities should be undertaken with utmost care for getting the desired results. Canal lining should be of high quality for its proper functioning to reduce seepage, and therefore should be done under technical supervision of engineers. Land leveling is crucial for large holdings for reducing field runoff and percolation losses (of applied water), and enhancing distribution uniformity.

Conclusion

NREGS, one of the largest social protection initiatives in the world, promised a lot on the water management front, along with rural employment. But a quick analysis shows that the employment created through NREGA, the main purpose the scheme was intended for, is very poor, in comparison to the target. The reasons are many. First, the employment-generation potential of rural areas through creation of public works is much lower in many states, when compared to the funds allocated to those states. Second, in many states, unemployed people actually available for doing wage labor under public works are much fewer in comparison to the number of job card holders, whereas the fund allocation is far in excess of the genuine needs of the states. Third, in some situations, while enough numbers of the unemployed population are available in villages, not all of them have sufficient incentive to participate in public works, due to the high opportunity costs. Either the wage rate offered is low or the scheme is not able to create as many days of employment as they can get through migration.

On the other hand, analysis of the types of interventions in the face of the macro hydrological, geological, geo-hydrological and topographical realities and limited field evidence suggests that the planning and implementation of WM works are seriously flawed due to the total absence of hydrological and economic analysis. The types of interventions chosen for execution in different zones are not based on considerations of agro-climate, hydrological regime, topography and geological settings, which are paramount in deciding the effectiveness of land- and water-based interventions. Further, there is a terrible shortage of key administrative and technical staff required for proper implementation of the scheme. Also, the capacity of rural local governments in performing their intended functions is weak, barring very few states like Kerala. Given these realities, it is important to devise long-term strategies for the implementation of works especially those related to water management, which among other things should include technical and managerial capacity building of local panchayats.

We have discussed the broader water-management strategies suitable for each one of the identified typologies. That said, within each typology, a lot of scientific inputs would be required for technical planning of the works suitable for any given locality. Some of this may have to come from detailed investigations of catchment hydrology, geo-hydrology and topography.

As evidence emerging from different parts of the country suggests, the belief that creating more storage space for runoff water in villages through digging tanks and ponds would help water conservation and management is dangerous, and can have several negative welfare effects because of the adverse social and ecological consequences this creates in the downstream areas. This can be in the form of drying up of drinking water tanks/ponds in villages (Bachelor et al., 2002), reduced flows into large reservoirs meant for irrigation and drinking (Kumar et al., 2006; Kumar et al., 2008), reduced recharge of downstream wells (Ray and Bijarnia, 2006) and excessive siltation of traditional water bodies. Thus, there is a need for careful planning and efficient implementation of WM works in order to make NREGS an effective social protection initiative capable of reducing rural poverty and enhancing livelihoods. But, it is imperative that when public funds to the tune of hundreds of billions of rupees are spent on creating assets in villages, a small fraction of it is spent for their scientific planning, and for local capacity building to execute them.

Notes

1. The title of the paper is adapted from the article titled "Crores of MNREGA Money Spent on Ghost Workers and Invisible Dams," published in *Times of India* on June 14, 2011.
2. In October 2009, it was rechristened as Mahatma Gandhi National Rural Employment Guarantee Act.
3. The budgetary allocation for NREGA was increased to Rs. 40,000 crore for 2011–12.
4. Millions of rural people from Bihar and central India migrate to states such as Punjab, Maharashtra, Haryana, and Kerala during the agricultural season to work on the farms and thus are not available for doing public works.
5. For instance, while the authors claimed the use of hydrological models for analyzing the impact on groundwater, there was no description of the hydrological model, which was supposed to have been attempted, in the paper. While PRA and hydrological monitoring were claimed to have been used for measuring groundwater level, how these data were used to arrive at recharge figures is not known. So was the case with the estimation of loss of water through runoff. Here again, hydrological modeling was supposed to have been employed. But there is neither a description of the model nor the estimates.

References

Ambasta, P., Vijay Shankar, P. S. and Shah, M. (2008). Two Years of NREGA: The Road Ahead. *Economic and Political Weekly, 43*(8), 41–50.

Batchelor, C., Singh, A., Rama Mohan Rao, M. S. and Butterworth, J. (2002). *Mitigating the Potential Unintended Impacts of Water Harvesting.* Paper presented at the IWRA International Regional Symposium on Water for Human Survival, New Delhi, India, November 26–29.

Baron, J. (2009). Background: The Water Component of Ecosystem Services and in Human Well-Being Development Targets. In J. Baron (Ed.), *Rainwater Harvesting: A Lifeline for Human Well-Being* (pp. 4–13) Sweden: United Nations Environment Programme/ Stockholm Environment Institute.

Bassi, N. and Kumar, M. D. (2010). *NREGA and Rural Water Management in India: Improving the Welfare Effects.* Occasional Paper No. 3, Institute for Resource Analysis and Policy, Hyderabad.

CSE (2008). *An Assessment of the Performance of The National Rural Employment Guarantee Programme in Terms of Its Potential for Creation of Natural Wealth in India's Villages*. New Delhi: Centre for Science and Environment.

Farrington, J., Holmes, R. and Slater, R. (2007). *Linking Social Protection and the Productive Sectors*. Briefing Paper 28. London: Overseas Development Institute.

Government of India (2008). The National Rural Employment Guarantee Act 2005 (NREGA). *Operational Guidelines* (3rd ed.). New Delhi: Department of Rural Development, Ministry of Rural Development, Government of India.

Government of India (2010). Report on Employment and Unemployment Survey (2009–10). Chandigarh: Labour Bureau, Ministry of Labour and Employment, Government of India.

Government of India (2013). Land Use Statistics at A Glance (2001–2002 to 2010–2011). New Delhi: Directorate of Economics and Statistics, Department of Agriculture and Cooperation, Ministry of Agriculture, Government of India.

Gupta, A. K. (2009, October 31). Money Growth Policy for Bihar: Social Forestry Project Links Rural Employment With Growing Saplings. *Down To Earth*.

Gupta, S., Javed, A. and Datt, D. (2003). Economics of Flood Protection in India. *Natural Hazards, 28*(1), 199–210.

Gupta, S. and Sidhartha (2011, 29 June). Farm Wages Shoot Up 43%. *Times of India Delhi*, 11.

Hagen-Zanker, J., McCord, A., Holmes, R., Booker, F. and Molinari, E. (2011). *Systematic Review of the Impact of Employment Guarantee Schemes and Cash Transfers on the Poor*. London: Overseas Development Institute.

The Indian Express (2006, 19 June). Rajasthan, MP Face-Off over Chambal Water.

Kumar, M. D. (2009). *Water Management in India: What Works, What Doesn't?* New Delhi: India: Gyan Publishing House.

Kumar, M. D., Bassi, N., Sivamohan, M. V. K. and Niranjan, V. (2011). Employment Guarantee and Its Environmental Impact: Are the Claims Valid? *Economic and Political Weekly, 46*(34), 69–71.

Kumar, M. D., Ghosh, S., Patel, A., Singh, O. P. and Ravindranath, R. (2006). Rainwater Harvesting in India: Some Critical Issue for Basin Planning and Research. *Land Use and Water Resources Research, 6*, 1–17.

Kumar, M. D., Narayanamoorthy, A. and Singh, O. P. (2009). Groundwater Irrigation Versus Surface Irrigation. *Economic and Political Weekly, 44*(50), 72–73.

Kumar, M. D., Patel, A., Ravindranath, R. and Singh, O. P. (2008). Chasing a Mirage: Water Harvesting and Artificial Recharge in Naturally Water-Scarce Regions. *Economic and Political Weekly, 43*(35), 61–71.

Kumari, M. (2010). *NREGA: Towards Full Employment, Equality and Empowerment*. Paper presented at the 52nd Annual Conference of the Indian Society of Labour Economics (ISLE), Dharwad, Karnataka, December 17–19.

McCord, A. and Farrington, J. (2008). Digging Holes and Filling Them in Again? How Far Do Public Works Enhance Livelihoods? *Natural Resource Perspectives, 120*. London: Overseas Development Institute.

National Council of Applied Economic Research/Public Interest Foundation (2009). *NCAER-PIF Study on Evaluating Performance of National Rural Employment Guarantee Act*. New Delhi: National Council of Applied Economic Research.

Panagariya, A. (2009, September 19). More Bang for the Buck. *Times of India Delhi*, 12.

Pandit, C. (2009). Some Common Fallacies About Floods and Flood Management. *Current Science, 97*(7), 991–993.

Pisharoty, P. R. (1990). *Characteristics of Indian Rainfall. Monograph*. Ahmedabad: Physical Research Laboratories.

Ray, S. and Bijarnia, M. (2006). Upstream vs. Downstream: Groundwater Management and Rainwater Harvesting. *Economic and Political Weekly, 41*(23), 2375–2383.

Seckler, D. (1996). *The New Era of Water Resources Management: From "Dry" to "Wet" Water Savings*. Research Report 1. Colombo: International Irrigation Management Institute.

Shah, T. (2009, November 5). Money for Nothing: Cash Transfer Is a Flawed Alternative to NREGS. *Times of India Delhi,* 14.

Sharma, P. (2010, 13 June). Labour Pains: Punjab Farmers Reap Bitter NREGA Harvest. *Times of India Delhi,* 16.

Sharma, R. (2009). *National Rural Employment Guarantee Act*. Paper presented at the Observer Research Foundation (ORF) National Workshop on NREGA for Water Management, Delhi, India, October 30.

Sharma, S. (2011, June 14). Dead Men Working: The Ghost Dams Scam. *Times of India Delhi,* 9.

Singh, S. and Gill, S. S. (2010). *Impact of NREGA on Rural Employment in Punjab*. Paper presented at the 52nd Annual Conference of the Indian Society of Labour Economics (ISLE), Dharwad, Karnataka, December 17–19.

Sjoblom, D. and Farrington, J. (2008). *The Indian National Rural Employment Guarantee Act: Will It Reduce Poverty and Boost the Economy?* Project Briefing No. 7. London: Overseas Development Institute.

The Times of India (2010, 13 September). Systemic Reform to Root Out Corruption Still Needed. 16.

Tiwana, N. S., Jerath, N., Ladhar, S. S., Singh, G., Paul, R., Dua, D. K. and Parwana, H. K. (2007). *State of Environment: Punjab – 2007*. Punjab, India: Punjab State Council for Science & Technology.

Tiwari et al. (2011). MNREGA for Environmental Service Enhancement and Vulnerability Reduction: Rapid Appraisal in Chitradurga District, Karnataka. *Economic and Political Weekly, 46*(20), 39–47.

World Bank (2011). *Social Protection for a Changing India* (Vol. II, Ch. 4, p. 84). Washington, DC: Author.

4 Benefit-sharing mechanism for hydropower projects

Pointers for northeast India

Neena Rao

Introduction

Globally, water has been used for mechanical power since ancient times. However, water mills came much into vogue during the Industrial Revolution in Europe, and it was only in the second half of the 19th century that generation of electricity through water and its transmission became a practical possibility. In India, the first hydropower project was set up on the river Cauvery, which started transmitting electricity in 1902 to gold mines owned by the consortium of British Companies that were 90 miles away from the power station.

Until the 1970s, hydropower was considered as one of the major drivers of development world over. Pandit Jawaharlal Nehru lauded the big projects as temples of modern India. During this period, countries at various stages of development were looking at hydropower development as an option that facilitated use of natural resources for the growth and modernization of their economies.

However, water-resource projects, involving construction of large dams, come with huge environmental and social costs such as displacement of large numbers of people, excessive silt accumulation in reservoirs and submergence of vast areas of forests, causing ecological destruction. The social activist movement Narmada Bachao Andolan is a major milestone in the history of dam building in India. That the benefits from hydropower projects do not necessarily trickle down to the poorest became a growing concern leading to development of new paradigms with respect to benefit sharing from such projects. Long-term benefit sharing became a guiding principle, and broadening of the focus of hydropower projects from mere power generation to integrated water and land resource management began to be advocated as best practices in the hydropower sector.

Despite the several negative externalities associated with hydropower generation, it continues to remain the largest renewable energy contributor in the world. There are countries such as Brazil and Norway that remain completely dependent on hydropower generation to meet their energy needs even today. In India too, one can witness an increasing thrust toward hydropower development since the 1990s. Consequently, there is a simultaneous evolution of the concept of 'benefit sharing.' The northeast has nearly 39% of the estimated hydropower potential of the country, 98% of which remains untapped.

Recognizing the energy needs of the country and the untapped hydropower potential, the government of India had started harnessing its hydropower potential, in line with the global trend at that time. This is clear from the fact that in 1963, India's hydropower share in the energy mix was 50%. However, this declined to about 21% by 2010 (Saxena and Kumar, 2010).

In order to correct the hydrothermal power ratio to meet the grid requirements and peak power shortages, the government of India undertook certain measures and announced hydropower policy first in 1998 and thereafter in 2007. During this process, the standing committee on energy came out with a list of issues facing the hydropower sector and gave recommendations that shed some light on the eagerness to woo investment in this sector (see Standing Committee on Energy, 2005).

Hydropower development has been given high priority in the country's development plans during the last 2 decades. Strategically the northeast was given priority. This was because of the following: first, for the well-being of the people of the region and for its potential contribution to the national economy. Second, it was for fostering links and economic relations with neighboring countries.

Hydropower potential of the northeast region has mostly remained untapped until recently for various reasons such as the status of development of the grid systems in the country; availability of economic and accessible sites near to the load centers in the other regions of the country; low demand for power in these sparsely populated regions; and considerations regarding the impact of hydropower development on the livelihoods of the indigenous population, the river ecosystem and the safety of dams in this seismically active zone (Rao, 2006). Today, capacity addition through hydropower development in the northeast has become a high priority objective for the government. While the total capacity addition in NE grid was only 21 MW during the 11th 5-year plan period, it was 510 MW in Sikkim. From this, one can clearly see the fervor with which Sikkim was catching up with other states in terms of hydropower capacity addition during the 11th 5-year plan period, which coincides with the construction of the NHPC Teesta V.

Interestingly, the lead agency for hydropower development in this state is the state government of Sikkim; however, private sector involvement is being promoted aggressively. The government policies in India encourage participation of local communities in designing and implementation of a benefit-sharing mechanism in order to assure a flow of sustained stream of benefits to the project-affected communities. However, in reality it is observed that the benefits from the hydropower are largely enjoyed by the people (electricity users) located outside the project-affected area. An appropriate mechanism for distribution of the benefits is not yet developed.

Thus, there have been varied perceptions and views expressed for and against this move of the government for accelerated hydropower development. There have been several agitations and representations made to the government from different sections of the society, especially in the state of Sikkim.

The focus of this study is to arrive at recommendations for an optimal 'benefit-sharing mechanism' that takes care of the needs and aspirations of all stakeholders, especially the local communities. It intends to give policy recommendations for effective implementation of the benefit-sharing mechanism through identification of the enabling factors and constraints, and the users' perception on it.

Thus the first objective of this study was to assess the mechanism for benefit sharing and its implementation in hydropower projects in Sikkim. The study intends to figure out what works and what does not in terms of benefit-sharing mechanisms, and to offer recommendations to further strengthen the benefit-sharing mechanisms of hydropower projects.

The specific objectives are:

- To find out what works what does not, i.e. the enabling factors and constraints, and
- To understand and analyze the users' perceptions of this benefit-sharing mechanism.

Methodology

Toward this end, information and data was generated by applying various methods consisting of desktop study of existing government documents, study reports, research studies and information from other secondary sources and field

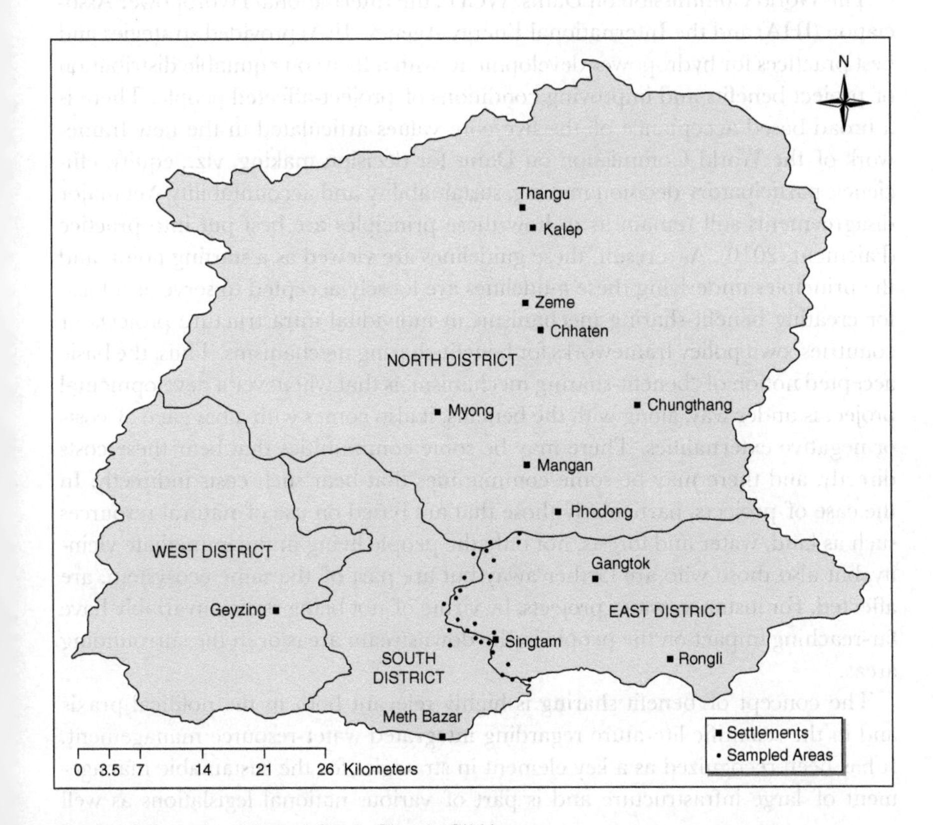

Figure 4.1 Location Map of Study Sites in Sikkim

study. The study was conducted in the following stages: reconnaissance visit, field visits, information from Geographical Information System (GIS), rapid assessment through focus group discussions and stakeholder workshops.The involvement of local stakeholders in information generation and use of their knowledge through interaction was an important component of the activities. The study was carried out in NHPC Teesta V project (MW 510) in Sikkim, India (please refer to Figure 4.1).

The concept of benefit sharing: Policy, legal and institutional aspects

It was in the 1980s that the concerns regarding the negative social and environmental impact of hydropower projects began to surface. But the issues of equity and sustainability became central to the debate on the benefit of large dams only in the 1990s, leading to emergence of new paradigms of benefit sharing. However, while the concept is often referred to, authors have often found it difficult to articulate what it means in reality. There is a dearth of conceptual and analytical framework to make the concept as useful as it could be to the practitioners.

The World Commission on Dams (WCD), the International Hydropower Association (IHA) and the International Energy Agency (IEA) provided strategies and best practices for hydropower development, with a focus on equitable distribution of project benefits and improving conditions of project-affected people. There is a broad-based acceptance of the five core values articulated in the new framework of the World Commission on Dams for decision making, viz., equity, efficiency, participatory decision making, sustainability and accountability. Yet major disagreements still remain as to how these principles are best put into practice (Paiement, 2010). As a result, these guidelines are viewed as a starting point, and the principles underlying these guidelines are loosely accepted to serve as a basis for creating benefit-sharing mechanisms in individual infrastructure projects or countries' own policy frameworks for benefit-sharing mechanisms. Thus, the basic accepted notion of 'benefit-sharing mechanism' is that whenever a developmental project is under way, along with the benefits, it also comes with a baggage of costs or negative externalities. There may be some communities that bear these costs directly, and there may be some communities that bear such costs indirectly. In the case of projects, particularly those that are based on use of natural resources such as land, water and forests, not only the people living in the immediate vicinity, but also those who are farther away but are part of the same ecosystem, are affected. For instance, water projects, by virtue of not being static, invariably have far-reaching impact on the people in the downstream areas or in the surrounding areas.

The concept of benefit sharing is highly relevant both in the political praxis and in the scientific literature regarding integrated water-resource management. It has been recognized as a key element in strategies for the sustainable management of large infrastructure and is part of various national legislations as well as in international compensation policies and guidelines (Bachurova, 2010).The principle is to share the benefits with communities that are affected directly and indirectly. Within this broad definition, two dimensions of benefit sharing can be differentiated: a) benefits for the locally affected communities, and b) trans-boundary benefit sharing (Bachurova, 2010).

Trans-boundary benefit sharing is based on the presumption that a common management of water resources generates net benefits compared to the unilateral development of the water resources. The concept is about the cooperation of riparian states for the use, protection or joint development of shared water bodies (trans-boundary rivers, lakes and aquifers), whereby the riparian states focus on the benefits from water cooperation and the win–win options instead of a potentially conflicting water sharing. ***Benefit sharing with the affected local population*** refers to a commitment to channel some of the returns generated by the operation of a project back to the population of municipalities, where water resources are exploited and infrastructure projects are developed. Apart from the classification of beneficiaries, the types of benefit can also broadly be divided into two types: monetary and nonmonetary (Bachurova, 2010).[1]

Many times hydropower projects are planned with the view of promoting equitable distribution of costs and benefits among different stakeholders in the backdrop of larger and sometimes even national-level realities. For instance, in India, it is a well-acknowledged fact that the country's energy needs require immediate attention. Therefore the benefits in such situations are viewed as national, whereas the costs to be borne are viewed largely as local and static in nature, i.e. one-time consequences. Planners often tend to neglect the long-term and far-reaching impacts/cumulative impacts due to the very structure of ecosystems, which is interconnectedness.

In view of the dire negative consequences of projects, the concept of benefit sharing focuses on long-term benefits. Its objective is to leverage long-term benefits for the affected communities. It is universally recognized that benefits should not be in the form of compensation alone, whether in cash or kind. Rehabilitation measures should also include steps that will restore and improve the livelihoods of those affected in the long run (Rossouw, 2010).

Regarding the form of benefits, i.e., whether the benefits should be distributed in kind or cash, Milewski et al. (1999) and Égré (2007) argue that benefits should be shared in monetary form only. However, this view has been criticized for being restrictive, considering the wide range of institutional challenges in developing countries. For example the National Rural Employment Guarantee Scheme, which is mired in corruption, has been a governance nightmare. The majority of these problems stem from the fact that cash payments do not effectively reach the intended beneficiaries. Also, monetary compensation and benefits can be temporary in nature. It is observed that there is a strong tendency amongst beneficiaries to spend off all that money for unproductive purposes. Hence there is a need to opt for a wide range of monetary and nonmonetary benefits (Rossouw, 2010).

However, who should be the beneficiary? What is the social and environmental damage, and to what extent should the project-affected people be compensated? And on what basis do they persist? This is due to improper assessment of indirect and intangible costs. This stems again from undue focus of planners and entrepreneurs on financial costs alone. The indirect costs include the trauma of whole communities being physically uprooted from their native villages; being relocated to alien areas; the introduction of whole new populations amidst them; and changes in ways of life and livelihoods, and socio-cultural support systems. Therefore, the quantum of compensation, form and method of delivery need to be refined further in order to address long-term well-being of the project-affected communities.

The World Commission on Dams supports the rights of the local communities as primary beneficiaries of the water-infrastructure projects (World Commission on Dams, 2000). The Commission stresses on rights of the local community as the foremost group for benefit sharing.

The WDC report provides a model of participatory decision making, the relevance of which goes far beyond the energy and water sectors. Its guidelines can be used by NGOs, people's movements and sympathetic professionals in the quest for transparency and democracy in decision-making processes, for community control over local resources, for social justice, environmental protection and the equitable and sustainable management of scarce resources. However, the WCD's guidelines are not recognized by international law, and its recommendations are not binding on any institution. It is up to the NGOs and people's movements to put pressure on governments, companies and funding institutions to comply with the WDC recommendations.

On the whole, from the above discussion, the following can be seen as the key enabling factors for benefit sharing mechanism:

- Policy and regulatory framework;
- Stakeholder engagement and communication;
- Partnership formation; and
- Institutions and capacity building.

If these four factors are firmly put in place and trust can be fostered, then the benefit-sharing mechanism is likely to evolve in a manner that addresses the legitimate needs of all stakeholders.

Benefit-sharing mechanism for NHPC TEESTA V

The benefit sharing-mechanism in the case of NHPC TEESTA operates mainly through the four following tools: namely, the royalty to the state government and local community; resettlement and rehabilitation policy; sustainable development policy; and activities under corporate social responsibility (Figure 4.2).

Most of these rehabilitation and resettlement policies are in kind or in cash compensations for loss of assets and livelihoods, while the services provided under sustainable development policy guidelines are services meant for maintaining the

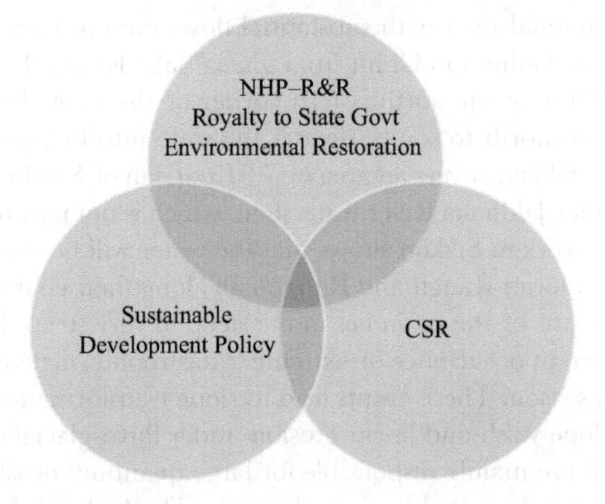

Figure 4.2 NHPC Benefit-Sharing Mechanism

health of the ecosystem and the services or activities undertaken under companies' CSR activities mainly related to general welfare of the project-affected community and the region. The concept of Payment for Ecosystem Services (PES) that is implemented elsewhere is not yet known to the people of Sikkim.

The concept of 'benefit-sharing mechanism' itself was new to the people of Sikkim. They were aware that they need to be compensated for the loss of their livelihoods and assets and that they were facing a lot of issues due to the changes brought about by hydropower development. However, they didn't seem to realize that apart from receiving compensation and some through corporate social responsibility: The community can demand a share in the long-term benefits from power developers. If the issue of 'lack of awareness' can be addressed by the government and power developers with a sensitive attitude, the project will be more sustainable and just.

The benefits received now are a result of a long drawn-out struggle and number of protests by the local communities. The record of the power developers with respect to compensation and compliance with environmental safeguards has been very poor. It is only during the last 4–5 years that the main public sector developer seems to have made significant progress in payment of compensation and compliance of environmental safeguards.

Part I: Characteristics of basin/sub-basin[2]

As a state that is a part of inner mountain ranges of Himalaya, Sikkim is entirely hilly, having no plain area with altitude varying from 213 m in the south to above 8,000 m in the northwest and north. The human habitable area is limited only up to the altitude of 2,100 m, constituting only 20% of the total area of the state.

Teesta is a perennial river with substantial flows even in lean seasons. Teesta River originates as Chhombo Chhu from glacial lake Khangchung Chho at an elevation of 5,280 m in the northeastern corner of the state. The main Teesta, while flowing from north to south, divides the state into two parts. The Teesta drainage basin in Sikkim covers an area of 7,015 sq. km of Sikkim, and 81 sq. km of the state is under Jaldhaka River watershed, which is not part of Teesta Basin. The drainage network in Sikkim shows that the water will be available in Teesta and its major tributaries Rangit and Rangpo all along their courses.

Since three-fourths of the basin lies in the steep to very steep slopes, it is characterized by recurrent occurrence of extreme (catastrophic) meteorological events during monsoon season. These events lead to slope transformation accompanied by gravitation, slope wash and linear erosion under fluvo-glacial environment in north Sikkim and are mainly responsible for large quantities of silt and aggradation material getting deposited in river channels. The high rainfall (about 2,300 mm) over the steeper slopes has created a suitable environment for initiation of runoff and subsequent soil erosion, slope failures, slides or sinking of land masses in Teesta Basin.

The carrying capacity report states that the areas that are most affected in the basin are the areas covered by softer rocks, viz., phyllites and schists of Daling Group. However, the areas where harder schist and even gneisses occur are also affected to a minor extent. Due attention must be given to the relationship between the attitude of litho units and road alignments. Landslide is one of the most pertinent and endemic problems, which is intensified manifold by human interferences, particularly in active mountainous regions like Sikkim Himalaya. The rainfall pattern in Sikkim Himalaya is observed to trigger landslides. The repeated thaw and freeze of the ice and the rainfall in the region have disrupted the roadways. The forests are also affected by landslide and erosion, which destroys valuable forestland and plantations and retards the growth of forest produce.

Forests constitute the major proportion of Teesta Basin in Sikkim and play an important role in maintaining the ecological balance and regulation of the hydrological regime of Teesta river system. In addition, these forests form the first resource of Sikkim and provide a wide range of forest-related services for the welfare of the human populace in Sikkim.

Water is a key environmental and economic resource for Sikkim. Any resource utilization, consumption and conservation must take into account both the ecological as well as economic criticality of this resource. The seven glacier complexes, which cover 17% of the area in the state, are the nodal points of water resource. Though Sikkim is located in one of the most human-populated regions in the country, it faces a shortage of drinking water.

The total drainage area of Teesta River Basin is15,240 sq. km. The catchment area of Teesta has quite a few glaciers in its upper reaches, the largest of which is the Zemu glacier in the western part of the catchment. The upper portion of the catchment area is reported to be covered with snow.

People's perceptions of the benefit-sharing mechanism

The respondents in our study carried mixed feelings about the royalty/benefits provided by the NHPC power projects in the different areas surveyed and found that benefits have percolated to those areas where NHPC colonies were located, or the areas where there is a high level of disturbance. Thus, uniform sharing of benefits to GPUs (wards) on a periodic time scale or frequency does not exist. All the benefits being provided to the beneficiaries as of now are need based and/or demand based.

Lower Samdong-Raley and Dikchu are the areas where the power developer NHPC has provided maximum benefits to the local community in the form of infrastructural development (construction of school, project hospitals, computer institute, health camps, training and capacity building, etc.) Interestingly, 57% of the upstream and downstream respondents said that they are unaware of the provision of royalty money and/or benefit-sharing mechanisms of the project developers. Only 43% of the sampled respondents know about the benefit-sharing programs (especially the CSR-CD programs) of the hydropower companies.

About 43% of the upstream/downstream respondents observed that hydropower developers had carried out the road construction and electrification in the project areas, in the NHPC colony in Raley and LANCO in Samardung Byasi, after repeated requests from the local communities.

Only 20% of respondents believe that business has flourished with the increase in the number of grocery/ration and stationery shops owing to the influx of the people (project workers) from the nearby states. Around 37% of respondents believe that the health and educational facilities have also improved after the project implementation. Furthermore, 37% of respondents are of the opinion that accessibility to market has increased with the construction of new roads. In the agricultural sector, only 3% of respondents believe that irrigation facilities have improved as a result of the project intervention. Around 17% of respondents feel that they have witnessed changes in occupational patterns of the people.

The sharing of benefits by NHPC Teesta Stage V among the communities is recorded since 2006. The criteria for the allocation of compensation or benefit sharing are both: 1) need based (contingent upon impact to households or landed property and/or impact due to natural disasters such as landslides and earthquakes) and 2) demand driven (from the communities/government agencies/local NGOs).

As per the testimony of NHPC officials, the monetary support/benefits provided by NHPC in 2011–12 are compensation for damage to households and private lands/farms, financial assistance for infrastructural development, contribution to chief ministers' relief funds for earthquake-affected people (for example, the September 18, 2011 quake) and allowances/ex-gratia (e.g. vibration allowance) to a few households.

Lower Samdong, Raley, Dipudara, Dikchu and Tumin in east Sikkim are the main areas where the benefits to the communities figure prominently. Nevertheless, the priority order of sharing or allocation has been changing over the years.

NHPC-V works in collaboration with the local NGOs (Yuva Jagriti Sangh, The Green Point) the member of legislative assembly (MLA) and the local government (gram panchayats) for the selection, planning and implementation of activities in the affected areas. However, there are several community-based organizations, formal and informal, that could be consulted and involved in the process. But only few organizations have been involved in the developmental activities of the projects.

NHPC has contributed around INR 140 crore annually to the state government revenue through 12% allocation of energy from the total power generation, NHPC officials said. We also met the HR official of the NHPC office at Balutar, who provided us with some literature about the activities.

Perceptions of the project-implementing agencies regarding the key issues

The project-implementing agencies raised issues about the Teesta River, which were different than that of the other stakeholders directly or indirectly affected by the hydropower projects. When the issue was discussed with the officials of NHPC Balutar, they argued that water availability or volume of water in the river during the lean season was the biggest issue. Seriously, the tension between the power developer and the local communities on the issue of acquiring land and employing local people in the project was another issue that emerged during the interviews with different project institutions. They also said that allocation/distribution of incentives/benefits among the affected people, their identification and development of benefit-sharing mechanisms posed a problem for them. Another issue that emerged from the interviews was the lack of adequate participation of local people, which largely hindered the effective implementation of developmental activities of the project. Almost all the project-implementing agencies shared this view.

Peoples' perception on public and private power developers and poor record of private sector in terms of benefit sharing

Among all the respondents, those who were favoring the public sector companies (NHPC) constituted the highest number. The survey pointed out that public sector companies are better than the private companies (e.g., LANCO, Sneha Kinetics) in the planning, selection and implementation of the community development programs and administration of royalty funds in the project areas. This can also be attributed to the participation of the local NGOs/CBOs in the developmental programs, which was higher in the public sector companies than with the latter.

The majority of the respondents (71%) argued that the public power developers such as NHPC are slightly better than private power developers. They further argued that public power developers provide at least some benefits to the communities in the form of royalty money and/or developmental activities while they (communities) haven't received any benefits/royalty money, etc., from the private power developers. Only a small fraction of the respondents (4%) were in favor of the latter. In addition, one-fifth of the respondents say that both public and private power companies have no major difference as such in the benefit-sharing process.

Table 4.1 List of Benefits Provided by the Public and Private Sector Companies (Based on the Respondents' Response)

Type of Benefit	*Location*	*Implementing Agency*	*Remarks*
Medical camps	Dikchu, Rashyap	NHPC, LANCO	Eye camps
Kendriya Vidyalaya	Balutar	NHPC	Sixty percent of students are from local areas
Road carpeting (black top)	Sing bel	NHPC	Reduced dust
A drinking water pipeline	Norgi Khola-Singbel	NHPC-Gammon	General public benefited
Afforestation	AlaichiKhola, Sakyong, Dochum, Doring, Lower Rakdong, Ralap	NHPC-FEWMD	Restoration of degraded areas
Bamboo plantation	Malbasey RF, Lower Namphing and Rashyap	NHPC/LANCO	Restoration of degraded areas
School bags and transportation for children	Sirwani School	LANCO	Poor children benefited
Steel suspension bridge	Singtam – connecting Singtam with Adarsh Busty and other towns of South Sikkim (on R. Teesta)	Government of Sikkim	Connects South Sikkim and East Sikkim at Singtam Bazar
Disaster relief (September 18 earthquake)	Earthquake-hit areas along the Singtam Dikchu stretch	NHPC	An amount of Rs. 3,000 was distributed to the affected households
Disaster relief	Affected households	NHPC	Rs. 5,000 each was given to few vibration affected households
Medical check-up	Dikchu, Adarsh Busty	NHPC/LANCO	Blood pressures/group, minor ailments etc.
Water channel	Upper Rashyap	NHPC	To control the rock/debris landslide triggered by LANCO project
Water cooler	Singtam hospital	NHPC	Drinking water for general public
Water filter	PHC Rangphu	NHPC	Drinking water for general public
Protection wall	Lower Rakdong	NHPC	Landslide protection measure

Source: Information based on field survey

Further, a small fraction of the respondents had no clear idea about the implementation of the benefit-sharing activities of those companies.

Perceptions regarding the role of public and private power developers

The majority of the respondents were reluctant to answer questions on preference for private or public developers, as they believed that impact from both the power developers is socio-culturally, socio-economically and socio-ecologically deleterious to the fragile landscapes of Sikkim. They also felt that all the rivers, lakes, falls, cliffs, mountain peaks, big trees, groves, springs/streams and small rivers are sacred, and they have traditional rituals to worship them for protection and prosperity (please refer to Figure 4.3).

However, on the whole, when asked, the majority of the respondents rated performance of public sector companies in benefit-sharing as better than that of private sector companies. Some of the reasons attributed are that the government looks at the investment in hydropower generation in remote areas like Sikkim as a very important role in the development of new schemes. However, the government recognizes its effort alone would not be sufficient to develop the vast hydropower potential. Therefore, it considers it essential to encourage greater private investment through IPPs and joint ventures by creating a conducive atmosphere for the private sector. However, considering the challenges specific to the hydropower sector, such as high cost, long gestation period and the high degree of uncertainty, the policy guidelines try to establish a level playing field for the private power developers by making the dispensation of regulated tariffs available to them all alike.

Communities' awareness of legal and policy issues

Only 42% of the respondents from the local institutions and the affected households were satisfied with the existing policy and legal provisions of royalty collection and administration by the power developers. On the other hand, 49% of respondents expressed dissatisfaction over the existing policy and legal provisions. They lamented that the existing policies and legal provisions of benefit sharing/

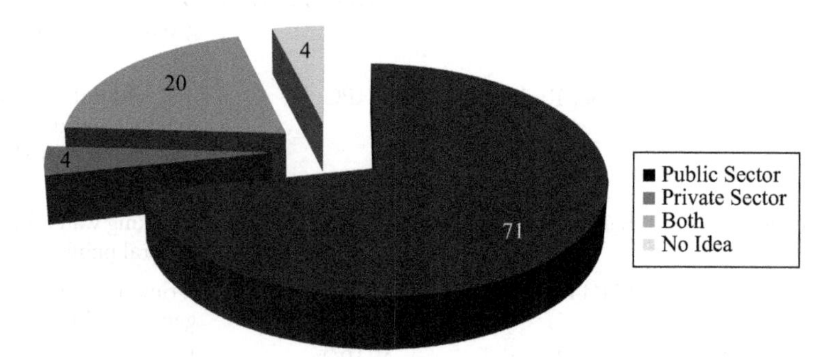

Figure 4.3 Proportion of the Respondents Favoring Public Sector Companies' Performance Over That of Private Sector Companies

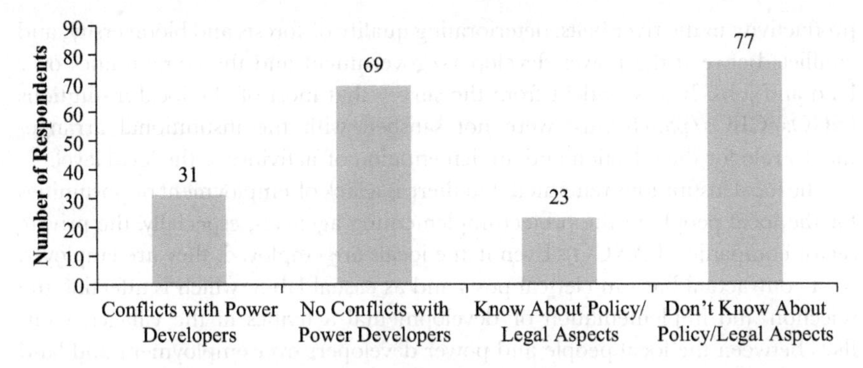

Figure 4.4 Respondents' Perceptions on Conflicts and Policies

allocation were hardly community oriented and demand based. They also reported that the allocated funds are not utilized evenly in all the project-affected areas but concentrated in the project-owned areas (see Figure 4.3).

Summary

One can see that the majority of the respondents except the project-implementing agencies, near unanimously, felt that there are serious concerns relating to ecological damage; flooding due to reservoir submergence; irregular water release through the Dikchu dam; drying up and disappearance of water bodies particularly rural springs and streams due to heavy tunneling landslides and soil erosion; and flash floods during monsoon. Sixty percent of the downstream respondents felt that the sand-quarrying business has been hampered due to non availability of sand resulting from irregular flow in the river. As a consequence, livelihoods of many families (wage laborers in quarries in the river banks), truck drivers and fishermen have been severely affected. The majority of the upstream and downstream stakeholders are against the dam projects, of which 70% opined that no project having slightest negative impact would be acceptable. Almost all the local institutional respondents felt that their participation in the decision-making process of NHPC Stage V was completely absent during project construction phase as well as after commissioning in 2008.

As regards health impacts, the respondents, especially officials from the health centers, expressed the concern that there was an emergence of new diseases due to heavy influx of population from the plains over the years into Sikkim. Cases of dengue fever, malaria, and HIV/AIDS have become common in Sikkim, they stressed. The district medical officer of Singtam expressed that the Singtam District Hospital is now unable to cater to the health services of the rapidly increasing population due to limited infrastructure and shortage of medical and paramedical staff. The other issues raised were pollution, traffic congestion and rising temperatures (less wind blowing due to low volume of water); lack of protection measures (wall) in reservoir area and along the river belt and human-inhabited area such as Manpari Busty in Singtam; landslides and flash floods; decline in agricultural productivity in the river belts; deteriorating quality of forests and biodiversity; and conflicts between the power developers/government and the communities over land and jobs. It was evident from the survey that most of the local institutions (NGOs/CBOs/panchayats) were not satisfied with the institutional arrangements/role for the selection and implementation of activities at the local level.

The local institutions reiterated that there is a lack of employment opportunities for the local people in the project-implementing agencies, especially, the private sector companies (LANCO). Even if the locals are employed, they are employed on a contractual basis in clerical posts and as casual labor, which is affecting the selection and implementation of developmental activities in the villages. Conflicts between the local people and power developers over employment and land acquisition were a common feature in the project-affected areas. Furthermore, the representatives of the local institutions indicated that constructions works of the power developers are mostly allocated to the nonlocal contractors.

The concerns raised during the field study were reiterated during the other key informant surveys and were also corroborated by desk research. For example, the information and scrutiny of official documents revealed the common safeguards stipulated by the Ministry of Environment and Forests, which are part of the MoU between the power developers and the state government,[3] were not put into practice. This was also supposed to be a 'Multidisciplinary Environmental Monitoring Committee (EMC)' constituted by the Ministry of Water Resources during February 1990 for overseeing the implementation of environmental safeguards stipulated by the Ministry of Environment and Forests while clearing the Water Resources Projects.[4]

However, in reality, compliance to these safeguards and actual monitoring by these monitoring committees, which were formed as a three-tier system, at the center, state and project level was very poor or almost negligible. For instance, in case of NHPC Teesta V the project was conceived in 1998, MoU was signed in 2000 with all the above-mentioned safeguards discussed in it; however, there is no record available of the monitoring committees meeting anytime before 2006.

The situation has not improved. It is evident from even after 2006, according to letters written by the monitoring committee members of the 'Gati Project' Monitoring Committee, which is operational in the area under this study.[5] Similar is the case with the records of the public hearings, where if the public hearings were held at all, they were not publicized well enough and consequently the list of attendees doesn't show much diversity in the participant stakeholders.

Conclusion and recommendations

To conclude, the perceptions of different stakeholders about the benefit-sharing mechanisms, impacts of projects and participation of communities in the decision-making process gives mixed understanding. If one looks at the development of hydropower from the country's energy needs, it is not difficult to understand the zest with which the governments are moving ahead to set up these power plants and wooing private investors for investment. The need for power generation is hardly a matter of debate. However, the way in which these power plants are being set up, the execution of these projects, right from the planning and design, to construction and operations, is a matter wanting attention.

The power developers in the public sector failed miserably in their implementation of benefit-sharing mechanisms and taking the local community into confidence in the initial phase. However, they fared better than the private players. The MoU between the government and NHPC for Teesta V was signed in the year 2000. It enlists all the environmental safeguards stipulated by the Ministry of Environment and Forests. However, there is no mention of the mechanisms for monitoring and evaluation to ensure compliance of these safeguards. The same was the case with the implementation of rehabilitation and resettlement policies.

Compliance with the environmental safeguards, public hearings and public consultations based on participatory approach were expected to be the integral part of the implementation process. However, in the initial phase of the project, none of these guidelines seem to have been followed properly. It was only after the people's uprising and frequent protests for their rights that action was begun in a positive direction.

A lot of benefits reached several locations in the project areas today. However, there are unserved locations that have not received their rightful benefits from the power developers as yet. These locations need more attention by the power projects. The upstream and downstream communities, in general, have voiced the concern that they still have several pending issues that the power projects have to take into consideration on a priority basis. Increased rate of pollution due to ever-increasing vehicular movement; traffic congestion; accidents; the impact of tunneling on local water bodies, such as springs; structural damage and cracks to private homes; land subsidence; and the fading of local culture and traditions were some of the impacts the stakeholders were concerned with.

The NHPC also claims to have brought about major changes in the communities by providing various incentives and benefits. They have contributed around Rs. 140 crore annually to the state revenue. The benefits have been in monetary form or as development activities or in kind, such as equipment/machines or infrastructure development. These claims need to be evaluated by way of a more comprehensive and in-depth study.

Nevertheless, NHPC has evolved its benefit-sharing mechanism considerably over a period of 10 years. It has also invested in the carrying capacity study of the Teesta River Basin, which cautioned that the basin could support only six mega hydropower projects. However, the Sikkim government went ahead, announcing at least 28 such hydropower projects to be developed by the private and public developers. Disturbingly, the record of private hydropower developers with respect to benefit sharing with the communities of projected-affected areas and environmental safeguard compliance has been dismal.

Notes

1. Monetary benefit sharing means sharing part of the monetary flows generated by the operation of the infrastructure project with the affected communities through the following: revenue sharing, preferential rates, property taxes, equity sharing/full ownership or development funds. Nonmonetary benefit sharing means integrating project benefits

into local development strategies such as livelihood restoration and enhancement, community development and catchment area development (see http://www.iwawaterwiki.org/xwiki/bin/view/Articles/BenefitSharingMechanisms).
2. The description of the physical and hydrological characteristics of the study area draws from the Carrying Capacity Report (Center for Interdisciplinary Studies of Mountain and Hill Environment, 2006) and the NHPC website (http://www.nhpc.co.in/index.htm).
3. These safeguards are as follows: a) drawing up a master plan for rehabilitation of the oustees; b) compensatory afforestation; c) restoration of construction areas; d) necessary arrangements for supply of fuel wood by the project authorities to the labor force during the construction period; e) to identify the critically eroded areas in the catchment for soil conservation work; f) mechanism for free movement of fish upstream and downstream of the structure across the river; g) setting up of monitoring units for implementing the suggested safeguards; h) alternatives in case of adverse effect on flora and fauna, wildlife etc.; and i) command area development.
4. The Committee is headed by the member (WP&P), CWC with director (environmental management) as its member/secretary. It comprises members from Ministries of Environment and Forests, Agriculture and Cooperation and Welfare and Water Resources, besides Planning Commission. The Committee is entrusted with the work of review of the mechanism established by the project authorities to monitor the ecology of the project areas and to suggest additional compensatory measures/facilities wherever necessary. It is also required to bring to the notice of the government important cases of default, which may lead to the review of project's clearance for the funding arrangement.
5. Based on testimony and letters to the monitoring committee provided by the head of an NGO in Sikkim who was one of the monitoring committee members.

References

Bachurova, A. (2010) Benefit Sharing Mechanisms. *IWA Water Wiki*. Downloaded from http://www.iwawaterwiki.org/xwiki/bin/view/Articles/BenefitSharingMechanisms

CISMHE (2006) Carrying Capacity Study of Teesta Basin in Sikkim. Introductory Volume, Vol. 1, University of Delhi, India

Égré, D. (2007) Benefit Sharing Issue, Compendium on Relevant Practices – 2nd Stage, UNEP Dams and Development Project, Nairobi, Kenya.

Milewski, J., Égré, D. and Roquet, V. (1999) Dams and Benefit Sharing, Contributing Paper prepared for Thematic Review I.1: Social Impacts of Dams, Equity and Distributional Issues, prepared as an input to the World Commission on Dams, Cape Town, www.dams.org.

Paiement, J. (2010) Hydropower and Affected Communities: International Experience with Benefit Sharing. Benefit Sharing Synthesis Report.

Rao, V. V. K. (2006) Hydropower in the Northeast: Potential and Harnessing Analysis. Background Paper No.6. Commissioned as an input to the study "Development and Growth in Northeast India: The Natural Resources, Water, and Environment Nexus", pp. 1–55.

Rossouw, N. (2010) Benefit Sharing with Local Impacted Communities During Project Implementation: Overview of the Literature and Practical Lessons from the Berg River Dam. Downloaded from http://www.cdri.org.kh/shdmekong/12.Berg_Benefit%20sharing.pdf

Saxena, P. and Kumar, A. (2010) Hydropower Development in India. Paper presented in The 8th International Conference on Hydraulic Efficiency Measures, IGHEM-2010, Oct. 21–23, 2010, AHEC, IIT Roorkee, India.

Standing Committee on Energy (2005) Thirteenth Report, Ministry of Non-Conventional Energy Resources, Lok Sabha Secretariat, New Delhi.

World Commission on Dams (2000) Dams and Development: A New Framework for Decision-Making, The Report of the World Commission on Dams. Earthscan: London.

5 Is irrigation development still relevant in reducing rural poverty in India?

An analysis of macro-level data

A. Narayanamoorthy and Susanto Kumar Beero

Introduction

Irrigation has been an important factor in the development of agriculture in India, which is clearly demonstrated by various studies (Gadgil, 1948; Dhawan, 1988; Rath and Mitra, 1989; Vaidyanathan et al., 1994; Narayanamoorthy and Deshpande, 2005). It helps not only to increase adoption of yield-increasing crop varieties, cropping intensity and productivity of crops but also to provide year-round employment opportunities and to push up the wage rates for agricultural laborers (Dhawan, 1991; Ray, 1992; Vaidyanathan, 1994; Narayanamoorthy and Deshpande, 2003). The impact of irrigation on poverty can be perceived through three different routes. First, the increased local availability and the resultant affordability of food grains due to lower prices reduce the number of poor people. Second, the increased labor absorption due to intensification of agriculture and consequent rise in wage rates and income make it possible for the rural poor to cross the poverty barriers. Third, the secondary effects of irrigation in terms of increased service opportunities within and outside the rural areas, better quality of life and consequent industrialization increase the tempo of economic activities in the irrigated regions, which results in poverty reduction. For instance, a study involving field research carried out in Andhra Pradesh found a dramatic change in cropping pattern in favor of sugarcane and banana due to increased availability of groundwater irrigation among both well-owning and non-well-owning farmers, which resulted in increased demand for labor and wage rate substantially (Shah and Raju, 1988). Both increased affordability of food grains and increased wage rates have substantially reduced poverty in areas where groundwater irrigation is well developed.

Although irrigation has been recognized as a crucial factor in explaining the growth in agriculture and rural economy in the literature, it has not been used as one of the principal explanatory variables in the poverty-related studies in India until 2001, when a study by Narayanamoorthy (2001) directly related the irrigation variable with rural poverty through a simple regression analaysis.[1] The explanation of cross-section variations in rural poverty has all along been sought through agricultural growth (Ahluwalia, 1978), average value of agricultural output per hectare (Sundaram and Tendulkar, 1988) or through output per hectare

(Datt and Ravallion, 1996; Ravallion and Datt, 1996). However, to the best of our knowledge, the use of irrigation as an explanatory variable has not been attempted by studies on rural poverty till the year 2001. After the publication of Narayanamoorthy (2001), quite a few studies have attempted to study the relationship between the irrigation development and rural poverty nexus using aggregated and disaggregated data in India and in South Asia (Saleth et al., 2003; Bhattarai and Narayanamoorthy, 2003; Hussain and Hanjra, 2003, 2004; Shah and Singh, 2004; Hussain, 2007a, 2007b, 2007c; Lipton, 2007; Narayanamoorthy, 2007; Narayanamoorthy and Hanjra, 2010). It is well recognized by these studies that irrigation is one of the principal causative factors behind the temporal changes in agricultural growth, and hence directly using irrigation as an explanatory factor is necessary. However, the available studies on the irrigation-poverty nexus have certain limitations. First, the variable irrigation has not been properly defined and used in the analysis; in most cases percentage of irrigated area to cropped area is used for the analysis, which does not capture the density of irrigation in relation to rural population. Second, irrigation has not been used as a lagged variable in the regression analysis, which is needed because irrigation cannot instantaneously impact on rural poverty. Third, most studies have used very old data covering the period only up to 1993–94, which does not capture the current reality. Fourth, studies covering all the major states with all the eight time points of poverty data, namely, 1973–74, 1977–78, 1983, 1987–88, 1993–94, 1999–2000, 2004–05 and 2009–10, are seldom available. Since many unprecedented changes have taken place in the rural areas since early nineties,[2] there is a need to study the poverty-irrigation relationship covering very recent data. Keeping this in view, an attempt has been made in this study to test the relationship between the incidence of rural poverty and the development of irrigation across the major states in India.

A snapshot of poverty-related studies

Since the present study aims to analyze the relationship between irrigation development and rural poverty over the years, it is better to understand the various studies available in this field of research before getting into the analysis. Quite a few studies have attempted to analyze the level of poverty for different periods in India. These poverty-related studies can be grouped into two broad categories, namely: a) the studies involving measurement of poverty (Ojha, 1970; Dandekar and Rath, 1971a, 1971b; Bardhan, 1973; Tyagi, 1982; Rath, 1996); and b) the studies analyzing the factors determining the incidence of poverty (Sundaram and Tendulkar, 1988; Dev, 1988, 1995; Kakwani and Subbarao, 1990; Nayyar, 1991; Singh and Binswanger, 1993; Ghosh, 1993,1996; Dasgupta, 1995; Vyas and Bhargava, 1995; Parthasarathy, 1995; Sharma, 1995; Hirway, 1995; Sagar, 1995; Bhalla, 1995; Kannan, 1995; Tendulkar and Jain, 1996; Datt and Ravallion, 1996). Before the study of Ahluwalia (1978), most studies dealt mainly with the estimates of rural and urban poverty and methodology for such estimation. Ahluwalia's (1978) study was one among the first attempts in explaining the variation in rural poverty in India.

For explaining the variations in the level of poverty, so far researchers have generally considered variables such as per capita consumption expenditure, per capita availability of food grains, consumer price index of agricultural laborers, state domestic product of agriculture per head of rural population, productivity of agricultural laborers, land holding size of different class of farmers, income of the agricultural labor households, number of days worked by the agricultural laborers and real wage rate. Similarly, land-man ratio, productivity of rain-fed land, inequality in land distribution, incidence of wage labor, average value of assets owned per household and extent of diversification in rural employment were also used as explanatory factors while analyzing poverty. These are, of course, interconnected in quite a complex process in impacting rural poverty. Therefore, it is rather difficult to map such process easily (Sundaram and Tendulkar, 1988). The above studies have arrived at two broad conclusions connecting to the incidence of poverty. While some studies have asserted an inverse relationship between agricultural growth and the incidence of rural poverty (Ahluwalia, 1978; Saith 1981; Ghosh 1993, 1996, 1998), others have raised some doubts about the existence of 'trickle-down process' in India (Bardhan, 1984, 1986; Mundle, 1983).

The present study does not aim at examining the validity of trickle-down hypothesis in India, but investigating the role of irrigation as an explanatory variable in rural poverty. There are two main reasons why irrigation should be considered as an important explanatory variable. First, since the irrigation facility through its production-augmenting and wage-enhancing effects substantially improves the flow of income, the inclusion of irrigation as an explanatory factor becomes very essential. Second, almost all the variables used by the earlier studies for testing the 'trickle-down process' are in one way or the other connected or determined by the level of irrigation. It is important to underscore some of these variables, which have high sensitivity to the availability of irrigation.[3] As mentioned earlier, the interconnection between these variables goes through a complex process. While it is essential to bring out the role of irrigation in explaining rural poverty at a point of time in a strict static sense, the dynamic aspect of such relationship has larger policy relevance. The questions like the long-run relationship of irrigation and poverty will give rise to the asymptotic behavior of the poverty-irrigation curve,[4] tapering off on the irrigation axis and creating a poverty trap explainable only through variables other than irrigation. In order to attempt this, we have tried to locate the changing relevance of irrigation as an explanatory factor for changing poverty rates over a period of time. The research question being addressed is whether irrigation is becoming increasingly less important as a determinant of poverty. The specific objectives of the study are: 1) to demonstrate the importance of irrigation as an impacting policy intervention on other variables that were used by the earlier studies for analyzing the incidence of rural poverty, 2) to analyze the changing scenario of rural poverty across the states from 1973–74 to 2009–10, 3) to analyze the relationship between the level of rural poverty and irrigation across the states over the years and 4) to find out whether or not the role of irrigation in reducing the rural poverty is improving over the years in India.

Data and methodology

The study aims to capture the variations in the incidence of rural poverty across the states, and therefore we have covered 14 major states of India for the analysis.[5] These 14 major states altogether accounted for about 95% of the total population below the poverty line in rural India both in the early seventies (1973–74) and also in the second decade of post liberalization (2009–10). The percentage of rural population below the poverty line released by the Planning Commission for eight different time points, namely 1973–74, 1977–78, 1983, 1987–88, 1993–94, 1999–2000, 2004–05 and 2009–10, has been used as the main data source for this study (Planning Commission, 2011). The other variables considered for the analysis are statewide data relating to percentage of gross irrigated area to gross cropped area (GIA/GCA),[6] irrigated area (ha) per thousand rural population (IAPTRP), real wage rate (Rs.) of agricultural laborers (RWAL), productivity (kg) of food grains (PFG), state domestic product (Rs.) of agriculture per head of rural population (SDAPHRP), cropping intensity (CI) and food grains production (kg) per head of rural population (FPPHRP). These data have been compiled and also estimated from the various issues of Indian Agricultural Statistics, Area and Production of Principal Crops in India (both are published by the Ministry of Agriculture, Government of India, New Delhi), Census of India, Fertilizer Statistics (published by the Fertilizer Association of India) and also from some of the recent published materials. The variables used in the study including their overall averages, etc., are presented in Table 5.1.

One of the objectives of the study is to demonstrate the impact of irrigation in other poverty-determining variables used by the earlier studies. For this, the following simple linear regression analysis (OLS method) is performed treating IAPTRP as an explanatory variable and other poverty-determining variables as dependent variables. This analysis is expected to reveal how closely the poverty-related variables used by the earlier studies are associated with the development of irrigation. In a way, the results arrived at from the following equation (1) would reinforce the need for considering irrigation development while studying the variations in rural poverty in India.

$$Y = \alpha + b_1 \text{ IAPTRP} \qquad (1)$$

[Where, Y = dependent variable used by the earlier studies; IAPTRP = irrigated area per thousand rural population; b_1 = regression parameters to be estimated and α = constant.]

As a second step of our analysis, states are divided into two groups based on the level of poverty, as states with above national average in rural poverty (ANA) and states with below national average in rural poverty (BNA). This is done to understand the trends in rural poverty and irrigation across different states. Apart from studying the nexus between irrigation development and rural poverty, an attempt is also made to study the trends in the degree of relationship between the two, i.e., whether the relationship between the two is weakening due to changes in other determining factors. In order to find out this relationship, keeping PRP

Table 5.1 Variables Used in the Study for Analysis

Variable	*Description of the Variables*	*1973–74 to 2009–10*				
		Unit	*Avg.*	*SD*	*Max.*	*Min.*
CI	Cropping Intensity	%	134.13	21.34	189.40	106.35
FPPHRP	Food Grains Production per Head Rural Population	kg	336.15	323.86	1,697.29	26.25
GIAGCA	Percentage of Gross Irrigated Area to Gross Cropped Area	ha	38.48	23.32	98.00	8.17
IAPTRP	Irrigated Area per Thousand Rural Population	ha	93.69	65.20	297.29	11.02
PFG	Productivity of Food Grains	g/ha	1,518.63	797.39	4,144.00	490.00
PRP	Percentage of Rural Poverty	%	36.31	17.08	73.16	6.35
RWAL	Real Wage Rate of Agriculture Laborer at 1986–87 Prices	Rs./day	16.86	9.88	66.32	4.92
SDAPHRP	State Domestic Product of Agriculture per Head of Rural Population at 1986–87 Prices	Rs.	1,085.90	854.16	4,880.23	149.93

Notes: SD: Standard Deviation; Avg: Average; ha: Hectare; kg: Kilogram; Max: Maximum; Min: Minimum.

Sources: Computed from Census of India (various years); Planning Commission (2011); www.planningcommission.nic.in; GOI (2012); GOI (various years).

as the dependent variable, three different types of simple linear regressions (OLS method) are computed, treating IAPTRP as an explanatory variable without any time lag and with 5 and 10 years' time lag for all time points using the following equations:

$$PRP_t = \alpha + b_1\, IAPTRP_t \tag{2}$$

$$PRP_t = \alpha + b_1\, IAPTRP_{t-5} \tag{3}$$

$$PRP_t = \alpha + b_1\, IAPTRP_{t-10} \tag{4}$$

[Where, PRP_t = percent of rural poverty in time t; IAPTRP = irrigated area per thousand rural population in time t,/in time $_{t-5}$ and in time $_{t-10}$; b_1 = regression parameter to be estimated and α = constant.]

Irrigation impact on other rural poverty-related variables

As a first step of our analysis, we try to demonstrate the impact of irrigation on other variables that were used by the earlier studies for analyzing the incidence of poverty in rural India. Variables like RWAL, SDAPHRP, FPPHRP, consumer price index, etc., have been mainly used by the earlier studies to study the relationship between the incidence of poverty and the growth of agriculture. There are no problems in using these variables to study the incidence of rural poverty because these variables determine the level of poverty. But these variables at best can be treated as the second layer of impact variables dictated by the availability of irrigation. The reason for considering them as secondary variables to irrigation is that these variables cannot act independently and are highly influenced by irrigation and water availability. In case the level of irrigation declines due to monsoon failures, it can cause an adverse impact on these variables. This can be easily understood through the analysis of the impact of irrigation on these variables empirically using the data of 14 major states of India. For this, we have computed correlation and regression using both traditional and nontraditional variables[7] (IAPTRP, RWAL, PFG, SDAPHRP, CI and FPPHRP) for all eight points of time considered for studying the incidence of poverty. However, since regression results give both the strength of association and its magnitude, we have presented only the results of regression computed treating IAPTRP as an independent variable to show the impact of irrigation on the other variables (see Table 5.2).

We will first explain the impact of irrigation on each of the traditional variables considered for the analysis. RWAL has been considered as one of the important variables for studying the incidence of rural poverty by some earlier studies (for example, Ghosh, 1996). State-level data show that this is positively influenced by the development of irrigation (Parthasarathy, 1996). The development of irrigation increases the intensity of cultivation, which in turn increases the demand for farm laborers. This process not only helps to increase the money wage rates and number of days of employment for agricultural laborers but also increases their total earnings. On the other hand, because of intensive cultivation, the production of agricultural commodities would increase, which results in reduction of price of essential commodities. This whole process ultimately helps to increase the real wage rates of agricultural laborers. This is very well corroborated by the results of our exercise, as the values of regression coefficients computed treating IAPTRP as an independent variable and RWAL as a dependent variable show that irrigation has a positive and significant relationship with RWAL at all the eight time points considered for the analysis.[8]

SDAPHRP is another important variable used by the earlier studies for studying the incidence of rural poverty. For instance, a study by Ghosh (1996) has used SDAPHRP as the only variable for analyzing the incidence of poverty across 14 major states of India for the years 1973–74, 1977–78, 1983 and 1987–88.

Table 5.2 Linear Regression Results: Impact of Irrigation on Other Related Variables

Year	*Dependent Variable*	*Model (1): Y = f (IAPTRP)*			
		Constant	*Slope*	R^2	*N*
1973–74					
	RWAL	6.05[a] (1.14)	0.04[a] (0.01)	0.54	14
	PFG	712.74[a] (149.82)	2.79[b] (1.32)	0.27	14
	SDAPHRP	507.93[a] (81.17)	4.26[a] (0.69)	0.76	14
	CI	111.96[a] (4.90)	0.12[a] (0.04)	0.38	14
	FPPHRP	95.41[a] (30.42)	2.06[a] (0.26)	0.83	14
1977–78					
	RWAL	7.03[a] (1.31)	0.05[a] (0.013)	0.45	14
	PFG	764.63[a] (168.56)	4.24[b] (1.62)	0.36	14
	SDAPHRP	−28.81 (33.19)	0.15[a] (0.04)	0.52	14
	CI	115.75[a] (6.22)	0.13[b] (0.06)	0.27	14
	FPPHRP	52.90[d] (33.23)	2.79[a] (0.32)	0.86	14
1983					
	RWAL	8.37[a] (1.52)	0.04[a] (0.02)	0.39	14
	PFG	789.55[a] (172.55)	5.18[a] (1.42)	0.53	14
	SDAPHRP	276.85[a] (73.74)	2.78[a] (0.61)	0.64	14
	CI	116.71[a] (7.17)	0.15[b] (0.05)	0.36	14
	FPPHRP	−0.51ns (40.49)	3.64[a] (0.34)	0.91	14
1987–88					
	RWAL	11.22[a] (2.05)	0.04[c] (0.02)	0.27	14
	PFG	692.05[a] (241.95)	7.49[a] (2.26)	0.48	14
	SDAPHRP	181.11[c] (90.97)	3.08[a] (0.85)	0.63	14
	CI	117.95[a] (7.79)	0.17[b] (0.07)	0.31	14
	FPPHRP	−42.39[b](56.76)	3.79[a] (0.53)	0.81	14
1993–94					
	RWAL	12.08[a] (2.67)	0.05[b] (0.02)	0.31	14
	PFG	992.59[a] (314.66)	6.89[b] (2.63)	0.37	14
	SDAPHRP	211.22[b] (85.28)	3.93[a] (0.71)	0.73	14
	CI	119.50[a] (8.78)	0.18[b] (0.07)	0.32	14
	FPPHRP	−112.22[d] (77.67)	4.85[a] (0.65)	0.83	14
1999–2000					
	RWAL	17.84[a] (4.52)	0.03 (0.04)	0.05	14
	PFG	955.74[b] (371.58)	9.78[a] (3.47)	0.34	14
	SDAPHRP	568.62[d] (358.36)	14.91[a] (3.34)	0.63	14
	CI	120.03[a] (10.59)	0.21[c] (0.09)	0.27	14
	FPPHRP	−189.96[c] (100.21)	6.05[a] (0.94)	0.78	14

(Continued)

Table 5.2 (Continued)

Year	*Dependent Variable*	*Model (1): Y = f (IAPTRP)*			
		Constant	*Slope*	R^2	*N*
2004–05					
	RWAL	24.71[a] (6.38)	–0.02 (0.05)	0.01	14
	PFG	1018.69[a] (360.67)	8.87[a] (3.19)	0.39	14
	SDAPHRP	548.58[c] (283.94)	14.19[a] (2.51)	0.73	14
	CI	121.54[a] (10.79)	0.20[b] (0.09)	0.29	14
	FPPHRP	–203.31[b] (89.25)	5.83[a] (0.79)	0.83	14
2009–10					
	RWAL	26.28[a] (7.34)	0.01(0.06)	0.002	14
	PFG	1308.89[a] (440.44)	6.95[c] (3.58)	0.24	14
	SDAPHRP	413.82[d] (241.53)	10.48[a] (1.97)	0.70	14
	CI	124.69[a] (12.75)	0.18[d] (0.11)	0.19	14
	FPPHRP	–213.42[d] (123.98)	5.78[a] (1.07)	0.74	14

Notes: a, b, c and d are significance level at 1, 5, 10 and 20% respectively; figures in parentheses are standard errors.

Sources: Computed from Census of India (various years); Planning Commission (2011); www.planningcommission.nic.in; GOI (2012); GOI (various years).

The study concluded that 'rural poverty is found to be inversely associated with agricultural production per head of rural population in all the time points' (p. 377). Nobody would dispute the fact that state domestic product of agriculture determines the level of rural poverty. However, it is important to investigate as to where the SDAPHRP is higher and what its determining factors are. One can definitely assert that SDAPHRP is mainly determined by the level of irrigation. Since the availability of irrigation determines the production and productivity of crops, the SDAPHRP of a state will be higher when the availability of irrigation is higher in any state.[9] This is also confirmed by the results of regression at all the points of time. Importantly, the regression results also show that the strength of the relationship between IAPTRP and SDAPHRP has become stronger between the early 1970s and late 2000s. For instance, while the slope of IAPTRP with respect to SDAPHRP was 4.147 in 1973–74, the same value increased to 10.48 in 2009–10. These results show that SDAPHRP is highly influenced by the level of irrigation (IAPTRP) in India.

The existing studies have also considered FPPHRP as one of the important variables to find out its impact on rural poverty. A few studies have recorded the inverse relationship between FPPHRP and rural poverty. As in the earlier case, here also the availability of irrigation determines the level of FPPHRP. It is quite obvious that in any state where the level of irrigation is higher, the production of food grains as well as the availability of food grains per head of rural population will also be higher. The regression computed treating IAPTRP as an independent variable (FPPHRP as dependent variable) precisely demonstrates this relationship. The value of regression coefficients has increased over the successive periods from 2.05 in 1973–74 to 3.79 in 1987–88 and further to 5.78 in 2009–10 (see Table 5.2). This result also undoubtedly confirms that irrigation is the main factor that determines the availability of food grains per head of rural population. Similar to this, the results of the variables such as cropping intensity (CI) and productivity of food grains (PFG) have also reflected their close relationship with the irrigation availability. Therefore, irrigation should be considered as primary variable in the analysis on rural poverty not only for its role as infrastructure per se but also for its inherent relationship with other determining variables of rural poverty.

Incidence of poverty and characteristics of the states

We have by now demonstrated how irrigation has significantly influenced the variables that are traditionally used in explaining the rural poverty by the existing studies. In this section, we try to identify the states that are above the national average (ANA) and below the national average (BNA) in terms of percentage of population below the poverty line and study their irrigation-related characteristics at all eight time points: 1973–74, 1977–78, 1983, 1987–88, 1993–94, 1999–2000, 2004–05 and 2009–10. The percentage of population living below poverty in rural areas has declined substantially at the national level (from 56.44% in 1973–74 to 33.80% in 2009–10) between the early 1970s and the end of the 2000s. However, the reduction is not the same across the states. While the rate of reduction is significant in states like Andhra Pradesh, Gujarat, Haryana, Karnataka, Madhya Pradesh, Kerala and West Bengal, it is less than the national level average in other states (see Table 5.3). With regard to individual states, despite a significant reduction in the level of poverty between 1973–74 and 2009–10, states like Orissa, Bihar, Madhya Pradesh and Uttar Pradesh continued to have a higher percentage of population below the poverty line in 2009–10. In all these states, the level of IAPTRP is lower than the national-level average in the remaining states. The percentage of rural poverty is not only less in the states of Punjab, Haryana, Gujarat, Rajasthan and Tamil Nadu but also declined significantly between 1973–74 and 2009–10, possibly because of higher IAPTRP. This suggests the important role played by IAPTRP in reducing the poverty in rural areas.

As the level of poverty is determined by many factors and their complex interactions, we have grouped the states into two groups: one with percentage of rural poverty above national average and the other with percentage of rural poverty below national average, for all eight time points to understand the unique characteristics of these groups. Here also we intend to show the importance of irrigation in determining the incidence of poverty. For analyzing the characteristics

Table 5.3 Data on Rural Poverty and Ranks of Each State, 1973–74 to 2009–10

States	*Rural Poverty (%)*							
	1973–74	*1977–78*	*1983*	*1987–88*	*1993–94*	*1999–2000*	*2004–05*	*2009–10*
1. Andhra Pradesh	48.41 (10)	38.11 (11)	26.53 (12)	20.92 (12)	15.92 (13)	11.05 (11)	11.20 (13)	22.80 (10)
2. Bihar	62.99 (3)	63.25 (4)	64.37 (2)	52.63 (2)	58.21 (1)	44.30 (2)	42.10 (2)	55.30 (1)
3. Gujarat	46.35 (11)	41.76 (10)	29.80 (11)	28.67 (11)	22.18 (12)	13.17 (10)	19.10 (9)	26.70 (7)
4. Haryana	34.23 (13)	27.73 (12)	20.56 (13)	16.22 (13)	28.02 (9)	8.27 (13)	13.60 (11)	18.60 (12)
5. Karnataka	55.14 (9)	48.18 (8)	36.33 (9)	32.82 (9)	29.88 (8)	17.38 (8)	20.80 (8)	26.10 (9)
6. Kerala	59.19 (5)	51.48 (7)	39.03 (8)	29.10 (10)	25.76 (11)	9.38 (12)	13.20 (12)	12.00 (14)
7. Madhya Pradesh	62.66 (4)	62.52 (5)	48.90 (5)	41.92 (5)	40.64 (5)	37.06 (3)	36.90 (3)	42.00 (2)
8. Maharashtra	57.71 (6)	63.97 (3)	45.23 (7)	40.78 (7)	37.93 (6)	23.72 (6)	29.60 (5)	29.50 (5)
9. Orissa	67.28 (2)	72.38 (1)	67.53 (1)	57.64 (1)	49.72 (2)	48.01 (1)	46.80 (10	39.20 (4)
10. Punjab	28.21 (14)	16.37 (14)	13.20 (14)	12.60 (14)	11.95 (14)	6.35 (14)	9.10 (14)	14.60 (13)
11. Rajasthan	44.76 (12)	35.89 (13)	33.50 (10)	33.21 (8)	26.46 (10)	13.74 (9)	18.70 (10)	26.40 (8)
12. Tamil Nadu	57.43 (7)	57.68 (6)	53.99 (4)	45.80 (4)	32.48 (7)	20.55 (7)	22.80 (7)	21.20 (11)
13. Uttar Pradesh	56.53 (8)	47.60 (9)	46.45 (6)	41.10 (6)	42.28 (3)	31.22 (5)	33.40 (4)	39.40 (3)
14. West Bengal	73.16 (1)	68.34 (2)	63.05 (3)	48.30 (3)	40.80 (4)	31.85 (4)	28.60 (6)	28.80 (6)
All India	56.44	53.07	45.65	39.09	37.27	27.09	28.30	33.80

Note: Figures in parentheses are 'Ranks.'

Sources: Planning Commission (2011) and www.planningcommission.nic.in.

of ANA and BNA states, we have taken into account all the variables (IAPTRP, RWAL, PFG, SDAPHRP, CI and FPPHRP), which are also considered in the earlier section.

The states that are coming under ANA group and BNA group in terms of percentage of poverty and their characteristics are presented in Table 5.4. Our

Table 5.4 States Having Above National Average (ANA) and Below National Average (BNA) in Rural Poverty and Their Characteristics

Year	*States*	*IAPTRP*	*RWAL*	*PFG*	*SDAPHRP*	*CI*	*FPPHRP*
1973–74							
ANA	Bihar, Kerala, Madhya Pradesh, Maharastra, Odisha, Tamil Nadu, Uttar Pradesh, West Bengal	55.44	8.20	940.00	703.69	123.26	207.99
BNA	Andhra Pradesh, Gujurat, Haryana, Karnataka, Punjab, Rajastan	136.84	10.90	998.33	1,171.48	121.75	378.88
1977–78							
ANA	Bihar, Madhya Pradesh, Odisha, Maharastra, Odisha, Tamil Nadu, West Bengal	52.74	8.25	1,000.00	698.44	126.00	232.97
BNA	Andhra Pradesh, Gujurat, Haryana, Karnataka, Kerala, Punjab, Rajasthan, Uttar Pradesh	106.63	12.50	1,207.50	831.18	126.56	325.56
1983							
ANA	Bihar, Madhya Pradesh, Odisha, Tamil Nadu, Uttar Pradesh, West Bengal	70.95	9.14	1,193.33	464.48	2,558.80	263.41
BNA	Andhra Pradesh, Gujarat, Haryana, Karanataka, Kerala, Maharashtra, Punjab, Rajastan	118.80	13.73	1,376.25	613.01	128.11	426.05
1987–88							
ANA	Bihar, Madhya Pradesh, Maharashtra, Odisha, Tamil Nadu, Uttar Pradesh, West Bengal	58.03	11.77	1,210.00	376.23	133.98	218.89
BNA	Andhra Pradesh, Gujarat, Haryana, Karanataka, Kerala, Punjab, Rajasthan	111.17	17.37	1,440.00	627.99	129.78	336.25

(Continued)

Table 5.4 (Continued)

Year	*States*	*IAPTRP*	*RWAL*	*PFG*	*SDAPHRP*	*CI*	*FPPHRP*
1993–94							
ANA	Bihar, Madhya Pradesh, Maharashtra, Odisha, Uttar Pradesh, West Bengal	70.86	14.28	1,435.00	462.46	140.04	281.98
BNA	Andhra Pradesh, Gujarat, Haryana, Karanataka, Kerala, Punjab, Rajasthan, Tamil Nadu	120.61	19.31	1,858.75	704.52	134.20	435.99
1999–2000							
ANA	Bihar, Madhya Pradesh, Odisha, Uttar Pradesh, West Bengal	67.48	16.14	1,639.20	1,206.27	146.08	253.69
BNA	Andhra Pradesh, Gujarat, Haryana, Karanataka, Kerala, Maharashtra, Punjab, Rajasthan, Tamil Nadu	105.60	23.25	1,974.67	2,346.76	134.90	428.58
2004–05							
ANA	Bihar, Madhya Pradesh, Maharashtra, Odisha, Uttar Pradesh, West Bengal	66.45	17.33	1,480.38	1,295.67	146.28	212.24
BNA	Andhra Pradesh, Gujarat, Haryana, Karanataka, Kerala, Punjab, Rajasthan, Tamil Nadu	117.10	27.94	2,152.65	2,358.34	138.94	457.66
2009–10							
ANA	Bihar, Madhya Pradesh, Odisha, Uttar Pradesh	93.21	18.67	1,612.00	933.13	149.05	259.55
BNA	Andhra Pradesh, Gujarat, Haryana, Karanataka, Kerala, Maharaashtra, Punjab, Rajasthan, Tamil Nadu, West Bengal	111.28	30.63	2,219.70	1,762.33	140.74	455.56

Sources: As in Table 5.2.

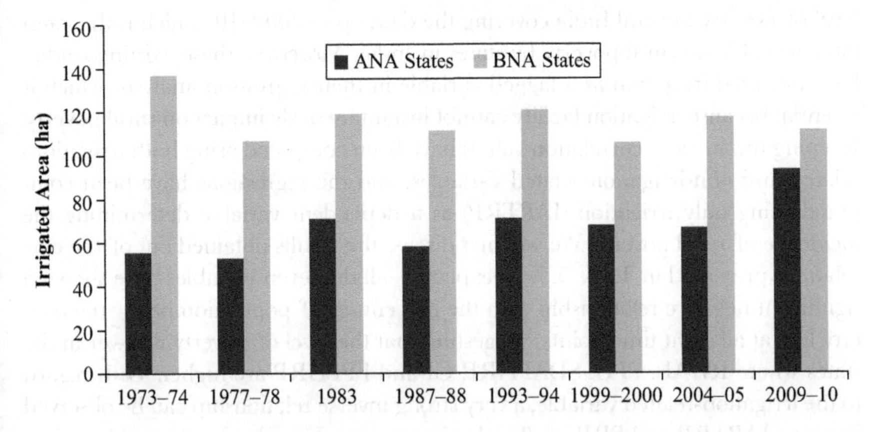

Figure 5.1 Irrigated Area per Thousand Rural Population for ANA and BNA Group of States in Terms of Rural Poverty

expectation here is that the BNA group of states will have significantly higher IAPTRP than the ANA group of states. As expected, IAPTRP is significantly higher for the group of BNA states when compared to the group of ANA states at all eight points of time considered for the analysis. While the average IAPTRP was 136.84 ha for the group of BNA states in 1973–74, it was only about 55.44 ha for the group of ANA states. Similarly, for the year 1987–88, while the average of IAPTRP was 111.17 ha for the group of BNA states, it was only 58.03 ha for the states belonging to the group of ANA states, showing a significant difference. This trend has been observed in the recent years too. For instance, the average IAPTRP was111.28 ha for the BNA states and only 93.21 ha for the group of ANA states for the year 2009–10.

This is found to be very much true for all other remaining time points taken for the analysis (see Figure 5.1). As far as other characteristics of the ANA and BNA states are concerned, as shown in the earlier section, the average values of RWAL, SDAPHRP and FPPHRP are much lower for the group of ANA states than for those of the BNA states group. All these suggest that irrigation plays a significant catalytic role in increasing the real wage rate for the agricultural laborers, SDAPHRP and FPPHRP. It needs to be mentioned here that all these variables were also considered as important determinants of poverty by the existing studies on rural poverty.

Irrigation and poverty nexus

As mentioned earlier, to the best of our knowledge, the study by Narayanamoorthy (2001) was probably the first study that directly related the irrigation factor with rural poverty using simple regression analysis, but this study has covered only four time points' data, namely 1973–74, 1977–78, 1983 and 1987–88. Since many unprecedented changes have taken place in agriculture and rural economy of India after 1987–88, this study tries to bring out the role of irrigation in determining the level of poverty in rural India covering the data up to 2009–10, which is the latest data available on rural poverty by states in India. Moreover, these existing studies have not used irrigation as a lagged variable in their regression analysis, which is essential because irrigation facility cannot instantaneously impact on rural poverty. Keeping this in view, correlation values have been computed using both irrigation-related and non-irrigation-related variables, and the regressions have been computed using only irrigation (IAPTRP) as a dependent variable determining the incidence of rural poverty. We will first discuss the results obtained out of the correlations presented in Table 5.5. As expected, all the seven variables have shown a significant negative relationship with the percentage of population below the poverty line at all eight time points, suggesting that the level of poverty is lower in the states where RWAL, PFG, SDAPHRP, CI and FPPHRP are higher. With regard to the irrigation-related variable, a very strong inverse relationship can be observed between IAPTRP and PRP at all eight time points. This clearly shows that irrigation is an important factor associated with the level of poverty in rural India.

The purpose of the study is not only to quantify the impact of irrigation on the rural poverty but also to examine whether the poverty-reducing impact of irrigation has increased or decreased over the years. In order to do this, we have computed three different regression models using ordinary least squares method involving IAPTRP as an explanatory variable with time lag of 5 years, with time lag of 10 years and also without any time lag for all eight time points. We are well aware of the fact that factors like RWAL, SDAPHRP, FPPHRP, etc., have significant influence on the level of poverty. However, we have excluded these variables while computing the regression equation mainly because of three reasons. First, our main aim here is to demonstrate the impact of irrigation on rural poverty and not to explain total variation in poverty across states. Second, as mentioned earlier, these variables are not the major independent variables because the level of irrigation highly influences these variables that are also already demonstrated adequately. Third, there is a possibility of multi-collinearity when these variables are included along with IAPTRP, as we

Table 5.5 Correlation Value: Percentage of Rural Poverty with Other Associated Variables

Variable	*1973–74*	*1977–78*	*1983*	*1987–88*	*1993–94*	*1999–2000*	*2004–05*	*2009–10*
GIAGCA	–0.626[a]	–0.653[a]	–0.513[c]	–0.527[b]	–0.301[ns]	–0.242[ns]	–0.207[ns]	–0.067[ns]
IAPTRP	–0.876[a]	–0.829[a]	–0.695[a]	–0.666[a]	–0.497[c]	–0.501[c]	–0.487[c]	–0.275[ns]
RWAL	–0.65[a]	–0.738[a]	–0.709[a]	–0.594[b]	–0.494[c]	–0.615[a]	–0.495[c]	–0.689[a]
PFG	–0.319[ns]	–0.466[c]	–0.494[c]	–0.485[c]	–0.384[d]	–0.45[c]	–0.592[b]	–0.56[b]
SDAPHRP	–0.818[a]	–0.478[c]	–0.448[d]	–0.65[a]	–0.603[b]	–0.65[a]	–0.706[a]	–0.601[b]
CI	–0.374[d]	–0.235[ns]	–0.362[ns]	–0.054[ns]	–0.011[ns]	–0.048[ns]	–0.094[ns]	–0.023[ns]
FPPHRP	–0.749[a]	–0.67[a]	–0.631[a]	–0.542[b]	–0.425[d]	–0.398[d]	–0.461[c]	–0.389[d]

Notes: a, b, c and d are significant level at 1, 5, 10 and 20% respectively.

Sources: As in Table 5.2.

have observed high correlation among these variables. Due to these reasons, we have computed regression using only irrigation (IAPTRP) as a single explanatory variable to relate with the level of poverty in rural India. The results of regression results estimated without giving any time lag for IAPTRP variable are presented in Table 5.6.

The regression results show that there is a significant inverse relationship between the level of rural poverty and the density of irrigated area standardized by rural population (irrigated area per thousand rural populations).[10] Though the results undoubtedly confirm that the development of irrigation reduces rural poverty, the strength of the regression coefficients along with R^2 has declined consistently from 1973–74 to 2009–10. For instance, while one hectare increase in IAPTRP reduced rural poverty by about 0.15% in 1973–74 and 0.131% in 1987–88, the extent of reduction was only 0.05% in 2009–10, when all 14 states were considered for the regression analysis. These results confirm our theoretical hypothesis of irrigation having a positive impact on rural poverty. However, such impact cannot be unidirectional in the sense that reduction in the pace of irrigation may not increase poverty or reduce the pace of poverty reduction. The reasons are obvious, and this can be explained with the help of growth in an agriculturally dependent population, relationship between agricultural and nonagricultural income and elasticity of such a relationship, as well as growth in real wages.

In order to study further the relationship between irrigation and the level of rural poverty, we have computed separate regression by eliminating West Bengal and Kerala. The reason for excluding these two states from the regression estimate is that despite low levels of irrigation development there, the poverty has declined significantly from 1972–73 to 2009–10. Moreover, as the agricultural laborers are the most vulnerable section of the rural society, the government of Kerala has introduced a social security scheme for this group. Under this scheme, pension is provided for the agricultural laborers above the age of 60, as well as for the physically disabled (Parthasarathy, 1996). A study conducted in this regard confirmed that as of 1986–87, one person out of every labor household got pension in Kerala (Kannan, 1990). Similarly, because of the pro-poor policies adopted in West Bengal over the years during the left rule, the rural poverty might have been reduced. As expected, the regression results computed after eliminating West Bengal and Kerala in the analysis showed a significant improvement in explaining the relationship between irrigation and rural poverty. The relationship between the two in terms of regression coefficients and R^2 has turned out to be much better as compared to the regression results computed without eliminating West Bengal and Kerala.

Development of irrigation cannot make a significant impact instantaneously on rural poverty in any given region. Irrigation benefit flows on certain pathways to finally make an impact on the poor people, which normally takes time. Irrigation availability initially changes the land-use pattern, including its intensity; increases the adoption of technological inputs and then brings changes in cropping patterns from low-value to high-value crops; improves the cropping

Table 5.6 Impact of Irrigation (IAPTRP) on Rural Poverty: Linear Regression Results

	Model (2): $PRP^t = f\ (LAPTRP^t)$								
	1973–74	*1977–78*	*1983*	*1987–88*	*1993–94*	*1999–2000*	*2004–05*	*2009–10*	*Pooled (1973–2010)*
Constant	67.58[a] (2.73)	67.32[a] (–4.28)	57.26[a] (5.64)	46.95[a] (–4.56)	42.07[a] (–5.52)	33.604[a] (6.43)	33.41[a] (5.34)	34.03[a] (6.18)	48.77[a] (2.45)
Slope	–0.152[a] (0.02)	–0.211[a] (0.04)	–0.155[a] (0.05)	–0.131[a] (0.04)	–0.091[c] (0.05)	–0.12[c] (0.06)	–0.091[c] (0.05)	–0.05[a] (0.05)	–0.133[a] (0.02)
R^2	0.77	0.69	0.49	0.45	0.25	0.26	0.24	0.075	0.26
N	14	14	14	14	14	14	14	14	112
Results arrived at excluding the data of Kerala and West Bengal									
Constant	66.34[a] (2.66)	68.99[a] (4.77)	58.84[a] (6.27)	49.39[a] (5.01)	45.93[a] (6.61)	40.06[a] (6.84)	38.98[a] (5.69)	42.43[a] (6.51)	52.12[a] (2.58)
Slope	–0.15[a] (0.02)	–0.23[a] (0.04)	–0.167[a] (0.05)	–0.15[a] (0.04)	–0.12[b] (0.05)	–0.16[b] (0.05)	–0.13[b] (0.05)	–0.11[c] (0.05)	–0.16[a] (0.02)
R^2	0.82	0.74	0.55	0.54	0.34	0.45	0.43	0.31	0.37
N	12	12	12	12	12	12	12	12	96

Notes: a, b, c and d are significance level at 1, 5, 10 and 20% respectively; figures in brackets are standard errors.

Sources: As in Table 5.2.

intensity; and then increases the production and productivity of crops. These changes increase not only the demand for labor but also the wage rate for agricultural laborers, among whom the incidence of poverty is very high in rural areas. The increased production of food grains owing to irrigation development also helps the rural poor to afford them at a cheaper rate, which ultimately helps them to cross poverty barriers. This entire process cannot take place instantaneously after the introduction of irrigation in any region. Therefore, it is necessary to use irrigation as a lagged variable in the regression analysis to capture its real impact on rural poverty. Keeping this in view, we have estimated regressions separately, treating irrigation with 5-years' time lag ($PRP_t = \alpha + b_1\ IAPTRP_{t-5}$) and also with 10 years' time lag ($PRP_t = \alpha + b_1\ IAPTRP_{t-10}$). The results show that the strength of impact of irrigation on reducing the rural poverty is much better when compared to the regression results computed treating irrigation without any time lag (see Tables 5.7 and 5.8). However, the strength of regression coefficients and the R^2 has been reducing over time even when irrigation is used as a lagged variable in the regression analysis. Does this mean that the irrigation is not very important in reducing rural poverty today in India? One cannot completely discard the relevance of irrigation in reducing rural poverty because even today when irrigation availability reduces due to monsoon failures

Table 5.7 Impact of Irrigation on Rural Poverty – Linear Regression Results Estimated Treating Irrigation With 5 Years' Time Lag

	Model (3): PRP = f (IAPTRP$_{t-5}$)							
	1973–74	*1977–78*	*1983*	*1987–88*	*1993–94*	*1999–2000*	*2004–05*	*2009–10*
Constant	67.48[a]	67.13[a]	57.21[a]	47.21[a]	40.72[a]	31.23[a]	33.96[a]	35.63[a]
	(3.04)	(3.99)	(5.58)	(4.70)	(5.06)	(6.14)	(5.57)	(5.60)
Slope	–0.17[a]	–0.19[a]	–0.18[a]	–0.12[a]	–0.09[c]	–0.08[d]	–0.10[c]	–0.07[d]
	(0.03)	(0.04)	(0.05)	(0.04)	(0.05)	(0.05)	(0.05)	(0.05)
R^2	0.73	0.72	0.48	0.45	0.24	0.19	0.24	0.15
N	14	14	14	14	14	14	14	14
Results arrived at excluding the data of Kerala and West Bengal								
Constant	66.14[a]	67.62[a]	58.81[a]	50.84[a]	43.15[a]	36.32[a]	40.42[a]	42.97[a]
	(2.99)	(4.41)	(6.29)	(5.89)	(5.95)	(6.95)	(5.91)	(5.42)
Slope	–0.16[a]	–0.19[a]	–0.17[a]	–0.14[c]	–0.11[b]	–0.12[b]	–0.15[a]	–0.12[b]
	(0.03)	(0.04)	(0.06)	(0.04)	(0.05)	(0.05)	(0.05)	(0.05)
R^2	0.77	0.75	0.55	0.57	0.31	0.33	0.45	0.43
N	12	12	12	12	12	12	12	12

Notes: a, b, c and d are significance level at 1, 5, 10 and 20% respectively; figures in brackets are standard errors.

Sources: As in Table 5.2.

Table 5.8 Impact of Irrigation on Rural Poverty – Linear Regression Results Estimated Treating Irrigation With 10 Years' Time Lag

	Model (4): PRP = f (IAPTRP$_{t-10}$)						
	1977–78	*1983*	*1987–88*	*1993–94*	*1999–2000*	*2004–05*	*2009–10*
Constant	67.14[a] (4.28)	57.09[a] (5.34)	47.75[a] (4.64)	41.97[a] (5.09)	30.31[a] (5.55)	31.97[a] (5.31)	35.62[a] (5.91)
Slope	–0.22[a] (0.04)	–0.17[a] (0.05)	–0.14[a] (0.04)	–0.09[b] (0.04)	–0.09[c] (0.05)	–0.07[d] (0.04)	–0.07[d] (0.05)
R^2	0.68	0.51	0.46	0.28	0.21	0.18	0.13
N	14	14	14	14	14	14	14
Results arrived at excluding the data of Kerala and West Bengal							
Constant	67.52[a] (4.71)	57.62[a] (6.91)	50.94[a] (5.05)	45.18[a] (5.89)	33.83[a] (6.14)	37.32[a] (5.96)	43.83[a] (5.77)
Slope	–0.22[a] (0.04)	–0.17[a] (0.05)	–0.17[a] (0.05)	–0.12[b] (0.05)	–0.12[b] (0.05)	–0.11[b] (0.05)	–0.14[b] (0.05)
R^2	0.72	0.56	0.57	0.38	0.32	0.35	0.42
N	12	12	12	12	12	12	12

Notes: a, b, c and d are significance level at 1, 5, 10 and 20% respectively; figures in brackets are standard errors.

Sources: As in Table 5.2.

or other reasons, the production of food grains and other agricultural commodities declines sharply, which results in increased food inflation, causing deep distress especially for the rural poor.

The reduced strength of regression coefficients of irrigation over time can be attributed to a few important factors. First, unlike in the 1960s, 1970s and 1980s, the rural non-agricultural income (earnings) has been growing faster than agricultural income especially for agricultural laborers during the last 2 decades, as confirmed by various studies (Vaidyanathan, 1994b; Visaria and Basant, 1994; Nayyar, 1996; Binswanger-Mkhize, 2013; Jatav and Sen, 2013). This possibly could have dampened the relationship between irrigation and rural poverty. Second, looking from another angle, the relationship might have depressed due to the intensity of various anti-poverty programs implemented in the late 1970s, increased by pace in the mid-1980s and in the 2000s. A quality change in the anti-poverty programs toward employment generation such as MGNREGS also could have made the difference. Probably, the strength of relationship between agricultural and non-agricultural income weakening over the years appropriately explains the dampening of the relationship between poverty and irrigation. On the whole, one can say from the above that the decline of a relationship between rural poverty and irrigation is possibly more due to significant increase of rural non-agricultural income than due to agricultural income during the last two decades.[11]

Conclusion and suggestions

Over the years, several attempts have been made by researchers for studying the causes for the rural poverty in India. Although irrigation plays an important role in reducing rural poverty, analysts have not explicitly brought out this idea. An attempt is made in this study to understand the role of irrigation in the reduction of the level of rural poverty in India, taking cross-section data of 14 major states at eight points of time: 1973–74, 1977–78, 1983, 1987–88, 1993–94, 1999–2000, 2004–05 and 2009–10. While analyzing the importance of irrigation empirically, the study shows through regression analyses that there is a clear significant inverse relationship between the incidence of rural poverty and the irrigated area per thousand rural people at all eight time points considered for the analysis. The study also finds that the strength of inverse relationship between irrigation availability and the rural poverty is better when irrigation is used as lagged variable, meaning that the base-level irrigation is also important in reducing rural poverty. However, the relationship between irrigation and rural poverty weakens over time, which could be possibly due to relatively faster growth of rural non-agricultural income than that of agricultural income during the last 2 decades. The massive steps taken by the government to alleviate rural poverty through various training and employment programs starting from the 1980s to the 2000s could also have weakened the relationship.

Despite the weakening relationship between irrigation and rural poverty, the study clearly establishes the existence of a significant inverse relationship between the two at all the eight time points considered for the analysis. The experiences from different regions in India also show that it is very difficult to achieve a sustained reduction in the level of rural poverty through anti-poverty programs, which have been implemented at a massive scale since the late 1970s. Rath (1996, p.106), who is one of the architects of India's poverty estimate, rightly mentions, 'in most states where today the incidence of poverty is high, irrigation is on the low side, with significant potentiality for its extension.' Therefore, judicious policies should be formulated to utilize untapped irrigation potential wherever possible on a priority basis, as the development of irrigation not only increases the production and productivity of agricultural commodities but also helps reduce the level of poverty through a percolation effect. The over-exploitation of groundwater that is taking place in semi-arid and arid regions in India can hamper the rate of reduction in poverty in rural areas by affecting the agricultural growth in the future. Therefore, policies need to be formulated to control this without harming agricultural growth.

Micro-irrigation systems, such as drip and sprinkler, are proven to be more efficient in irrigation water use, which needs to be aggressively promoted in those areas where over-exploitation of groundwater is serious to reduce rural poverty in a sustained manner. Minor irrigation (surface) sources, such as tanks, are very important for the resource-poor rural people in protecting their livelihood opportunities in many parts of India, particularly in southern India. But, unfortunately, the area under tank irrigation, which is the most user friendly and cheapest source of irrigation, has been declining consistently since the introduction of new agricultural technology in India due to negligence, poor maintenance and adverse effects of groundwater overdraft on tank hydrology. Therefore, efforts are needed to revitalize the small water bodies not only to improve the performance of agriculture but also to support the livelihood opportunities of poor people living below the poverty line. Improving water security should be the prime motto of the policymakers to have sustained reduction in rural poverty.

Notes

1. To the best of our knowledge, for the first time in poverty-related studies, Datt and Ravallion (1996) have used the initial irrigation rate (IRR) that is a percentage of operated area irrigated in 1957–60 as one of the variables while explaining poverty. But the IRR in the study was used more as an indicator of infrastructure.
2. Many unprecedented changes have taken place in rural India during the last 2 decades. Farmers' suicides, indebtedness, crop failures, unremunerative prices for crops and poor returns over cost of cultivation are the foremost features of India's agriculture today. Farmers committing suicide were not apparent before the early 1990s, but it has become a widespread phenomenon today in many states in India. Over two lac farmers have committed suicide in India between 1990–91 and 2009–10.
3. It needs to be reported here that real wage rate of agricultural laborers, per capita production of food grains to rural population, state domestic product (SDP) in agriculture per head of rural population, average working days of the agricultural laborers, etc., are mainly determined by the availability of irrigation facility. In this regard, a study (Parthasarathy, 1996) conducted, covering all the major states of India, has found a faster increase of real wage rate in the irrigated areas as compared to the unirrigated areas, suggesting the relevance in irrigation infrastructure for influencing wage rate.
4. The poverty-irrigation curve is defined by taking the rural poverty ratio on the y-axis and irrigation on the x-axis. Theoretically, we perceive that the curve will have an asymptotic with x-axis. In other words, it is difficult to perceive a point of zero poverty at any given level of irrigation.
5. The names of the states considered for this study are given in Table 5.3, along with their level of poverty.
6. We tried to ascertain the relationship of poverty with the ratio of irrigated area (GIA/GCA) as well as with IAPTRP. However, the ratio of irrigated area had a weaker relation in explaining the variation in rural poverty when compared with IAPTRP. Among the two, IAPTRP is more meaningful in explaining poverty, as this variable has population as base (standardized by population to eliminate that effect).
7. The variables such as RWAL, SDAPHRP, FPPHRP, etc., have been traditionally used by the studies, and therefore, we refer them as traditional variables. The variables introduced in this study such as IAPTRP, PFG and CI are treated as nontraditional variables.
8. The overall relationship is positive and significant between GIA/GCA and RWAL, as well as between IAPTRP and RWAL, but this is not true especially for Kerala and West Bengal. The real wage rate of agricultural laborers is higher in these two states despite a very low availability of irrigation per thousand rural population.
9. A recent comprehensive study carried out covering most districts of India has also showed a close relationship between the value of output per hectare (which is a proxy

variable to SDAPHRP) and the coverage of irrigation to cropped area. For more details on this, see Bhalla and Singh (2012).

10. Another set of regression equations computed taking percentage of gross irrigated area to gross cropped area as an explanatory variable also shows a significant inverse relationship with percentage of rural poverty. These results are not presented here for the purpose of brevity.
11. There has been a considerable increase in nonfarm rural income since the 1980s, which is a bright spot in India's rural economy. The growth of nonagricultural rural employment and its impact on earnings (income) and on rural poverty has been brought out by some studies. For more details in this regard, see Vaidyanathan (1994b); Visaria and Basant (1994) and Binswanger-Mkhize (2013).

References

Ahluwalia, M.S. (1978, April). Rural Poverty and Agricultural Performance in India. *Journal of Developmental Studies, 14*(2), 298–323.

Bardhan, P. K. (1973, February). On the Incidence of Poverty in Rural India of the Sixties. *Economic and Political Weekly, 8*(4–6), 245–254.

Bardhan, P. K. (1984). *Land, Labour and Rural Poverty*. Delhi: Oxford University Press.

Bardhan, P. K. (1986). Poverty and Trickle-Down in Rural India: A Quantitative Analysis. In J. W. Mellor and G. M. Desai (Eds.), *Agricultural Change and Rural Poverty: Variations on a Theme by Dharm Narain* (pp. 76–94). Delhi: Oxford University Press.

Bhalla, G. S. and Singh, G. (2012). *Economic Liberalization and Indian Agriculture: A District-level Study*. New Delhi: Sage Publications.

Bhalla, S. (1995, October 14–21). Development, Poverty and Policy: The Haryana Experience. *Economic and Political Weekly, 30*(41 and 42), 2619–2634.

Bhattarai, M. and Narayanamoorthy, A. (2003). Impact of Irrigation on Rural Poverty in India: An Aggregate Panel-Data Analysis. *Water Policy, 5*(5–6), 443–458.

Binswanger-Mkhize, H. P. (2013). The Stunted Structural Transformation of the Indian Economy: Agriculture, Manufacturing and the Rural Non-Farm Sector. *Economic and Political Weekly, 58*(26–27), 5–13.

Dandekar, V. M. and Rath, N. (1971a, January 2). Poverty in India – I: Dimensions and Trends. *Economic and Political Weekly, 6*(1), 25–48.

Dandekar, V. M. and Rath, N. (1971b, January 9). Poverty in India – II: Policies and Programmes. *Economic and Political Weekly, 6*(2), 106–145.

Dasgupta, B. (1995, October 14–21). Institutional Reforms and Poverty Alleviation in West Bengal. *Economic and Political Weekly, 30*(41–42), 2691–2702.

Datt, G. and Ravallion, M. (1996). *Why Have Some Indian States Done Better Than Others at Reducing Rural Poverty?* Policy Research Working Paper No. 1594. Washington, DC: Policy Research Department, Poverty and Human Resources Division, The World Bank.

Dev, M. S. (1988, January–March). Poverty of Agricultural Labour Households in India: A State Level Analysis. *Indian Journal of Agricultural Economics, 43*(1), 14–25.

Dev, M.S. (1995, October 14–21). Alleviating Poverty: Maharashtra Employment Guarantee Scheme. *Economic and Political Weekly, 30*(41 and 42), 2663–2676.

Dhawan, B. D. (1988). *Irrigation in India's Agricultural Development: Productivity, Stability, Equity*. New Delhi: Sage Publications.

Dhawan, B. D. (1991, October–December). Role of Irrigation in Raising Intensity of Cropping. *Journal of Indian School of Political Economy, 3*(4), 632–671.

Gadgil, D. R. (1948). *Economic Effects of Irrigation: Report of a Survey of the Direct and Indirect Benefits of the Godavari and Pravara Canals*. Publication No. 17. Pune, Maharashtra: Gokhale Institute of Politics.

Ghosh, M. (1993, April–June). Test of Trickle-Down Hypothesis in Rural West Bengal. *Indian Journal of Agricultural Economics, 48*(2), 216–225.

Ghosh, M. (1996, July–September). Agricultural Development and Rural Poverty in India. *Indian Journal of Agricultural Economics, 51*(3), 374–380.

Ghosh, M. (1998, November 2 – December 4). Agricultural Development, Agrarian Structure and Rural Poverty in West Bengal. *Economic and Political Weekly, 33*(47 and 48), 2987–2995.

GOI (1993, July). *Report of the Expert Group on Estimation of Proportion and Number of Poor*. New Delhi: Perspective Planning Division, Planning Commission.

GOI (1993a, September). *Area and Production of Principal Crops in India: 1990–93(and Various Years)*. New Delhi: Ministry of Agriculture.

GOI (1993b, September). *Indian Agricultural Statistics: 1985–86, 1989–90(and Various Years)* Vol. I (Summary Tables). New Delhi: Ministry of Agriculture.

Hirway, I. (1995, October 14–21). Selective Development and Widening Disparities in Gujarat. *Economic and Political Weekly, 30*(41 and 42), 2603–2618.

Hussain, I. (2007a). Direct and Indirect Benefits and Potential Disbenefits of Irrigation: Evidence and Lessons. *Irrigation and Drainage, 56*(2–3), 179–194.

Hussain, I. (2007b). Pro-Poor Intervention Strategies in Irrigated Agriculture in Asia: Issues, Lessons, Options and Guidelines. *Irrigation and Drainage, 56*(2–3), 119–126.

Hussain, I. (2007c). Poverty-Reducing Impacts of Irrigation: Evidence and Lessons. *Irrigation and Drainage, 56*(2–3), 147–164.

Hussain, I. and Hanjra, M. A. (2003). Does Irrigation Water Matter for Rural Poverty Alleviation? Evidence From South and South-East Asia. *Water Policy, 5*(5–6), 429–442.

Hussain, I. and Hanjra, M. A. (2004). Irrigation and Poverty Alleviation: Review of the Empirical Evidence. *Irrigation and Drainage, 53*(1), 1–15.

Jatav, M. and Sen, S. (2013). Drivers of Non-Farm Employment in Rural India: Evidence From the 2009–10 NSSO Round. *Economic and Political Weekly, 58*(26–27), 14–21.

Kakwani, N. and Subbarao, K. (1990, March 31). Rural Poverty and Its Alleviation in India. *Economic and Political Weekly, 25*(13), A2–A16.

Kannan, K. P. (1990). *State and Union Intervention in Rural Labour – A Study of Kerala, India*. ARTEP Working Papers, Asian Regional Team for Employment Promotion (ARTEP). New Delhi: International Labour Organization.

Kannan, K. P. (1995, October 14–21). Declining Incidence of Rural Poverty in Kerala. *Economic and Political Weekly, 30*(21 and 22), 2651–2662.

Lipton, M. (2007). Farm Water and Rural Poverty Reduction in Developing Asia. *Irrigation and Drainage, 56*(2–3), 127–146.

Mundle, S. (1983, June 26). Effect of Agricultural Production and Prices on Incidence of Rural Poverty: A Tentative Analysis of Inter-State Variations. *Economic and Political Weekly, 18*(26), A48–A61.

Narayanamoorthy, A. (2001, January–March). Irrigation and Rural Poverty Nexus: A State-wise Analysis. *Indian Journal of Agricultural Economics, 56*(1), 40–56.

Narayanamoorthy, A. (2007, April–July). Does Groundwater Irrigation Reduce Rural Poverty? Evidence From Indian States. *Irrigation and Drainage, 56*(2–3), 349–362.

Narayanamoorthy, A. and Deshpande, R. S. (2003). Irrigation Development and Agricultural Wages: An Analysis Across States. *Economic and Political Weekly, 38*(35), 3716–3722.

Narayanamoorthy, A. and Deshpande, R. S. (2005). *Where Water Seeps! Towards a New Phase in India's Irrigation Reforms*. New Delhi: Academic Foundation.

Narayanamoorthy, A. and Hanjra, M. A. (2010). What Contributes to Disparity in Rural-Urban Poverty in Tamil Nadu? A District Level Analysis. *Indian Journal of Agricultural Economics, 65*(2), 228–244.

Nayyar, R. (1991). *Rural Poverty in India, An Analysis of Inter-State Differences*. Bombay: Oxford University Press.

Nayar, R. (1996). New Initiatives for Poverty Alleviations in Rural India. In C. H. Hanumantha Rao and H. Linnemann (Eds.), *Economic Reforms and Poverty Alleviation in India* (pp. 171–198). New Delhi: Sage Publications.

Ojha, P. D. (1970, January). A Configuration of Indian Poverty: Inequality and Levels of Living. *Reserve Bank of India Bulletin, 24*(1), 16–27.

Parthasarathy, G. (1995, October 14–21). Public Intervention and Rural Poverty: Case of Non-Sustainable Reduction in Andhra Pradesh. *Economic and Political Weekly, 30*(41 and 42), 2573–2586.

Parthasarathy, G. (1996, January–June). Recent Trends in Wages and Employment of Agricultural Labour. *Indian Journal of Agricultural Economics, 51*(1 and 2), 145–167.

Planning Commission (2011). *Indian Planning Experience: A Statistical Profile*. New Delhi: Planning Commission, Government of India.

Rath, N. (1996, January–June). Poverty in India Revisited. *Indian Journal of Agricultural Economics, 51*(1 and 2), 76–108.

Rath, N. and Mitra, A. K. (1989, March). Economics of Irrigation in Water Scare Region: A Study of Maharashtra. *Artha Vijnana, 31*(1), 1–129.

Ravallion, M. and Datt, G. (1996, September). India's Checkered History in Fight Against Poverty: Are There Lessons for the Future? *Economic and Political Weekly, 31*(35–37), 2479–2485.

Ray, S. K. (1992, October–December). Development of Irrigation and Its Impact of Pattern of Land Use, Output Growth and Employment Generation. *Journal of Indian School of Political Economy, 4*(4), 677–700.

Sagar, V. (1995). Public Intervention for Poverty Alleviation in Harsh Agro-Climatic Environment: Case of Rajasthan. *Economic and Political Weekly, 30*(41–42), 2677–2690.

Saith, A. (1981, January). Production, Prices and Poverty in Rural India. *Journal of Developmental Studies, 17*(2), 196–213.

Saleth, R. M., Namara, R. and Samad, M. (2003). Dynamics of Irrigation-Poverty Linkages in Rural India: Analytical Framework and Empirical Analysis. *Water Policy, 5*(5–6), 459–473.

Shah, T. and Raju, K. V. (1988). Groundwater Markets and Small Farmer Development. *Economic and Political Weekly, 23*(13), A23–A28.

Shah, T. and Singh, O. P. (2004). Irrigation Development and Rural Poverty in Gujarat, India: A Disaggregate Analysis. *Water International, 29*(2), 167–177.

Sharma, N. A. (1995, October 14–21). Political Economy of Poverty in India. *Economic and Political Weekly*, 30(41 and 42), 2587–2602.

Singh, R. P. and Binswanger, H. P. (1993, January–March). Income Growth in Poor Dryland Areas of India's Semi-Arid Tropics. *Indian Journal of Agricultural Economics, 48*(1), 51–64.

Sundaram, K. and Tendulkar, S. D. (1988). Towards an Explanation of Interregional Variation in Poverty and Unemployment in Rural India. In T. N. Srinivasan and P. K. Bardhan (Eds.), *Rural Poverty in South Asia* (pp. 316–362). Delhi: Oxford University Press.

Tendulkar, S. D. and Jain, L. R. (1996, January–June). Growth, Distributional Change and Reduction in Rural Poverty Between 1983 and 1987–88. A Decomposition Exercise for Seventeen States in India. *Indian Journal of Agricultural Economics,51*(1 and 2), 109–133.

Tyagi, D. S. (1982, June 26). How Valid Are the Estimates of Trends in Rural Poverty? *Economic and Political Weekly, 17*(26), A54–A62.

Vaidyanathan, A. (1994, December 10). Employment Situation: Some Emerging Perspectives. *Economic and Political Weekly, 29*(50), 3147–3156.

Vaidyanathan, A., Kumar, A. K., Rajagopal, A. and Varatharajan, D. (1994, October–December). Impact of Irrigation on Productivity of Land. *Journal of Indian School of Political Economy, 6*(4), 601–645.

Visaria, P. and Basant, R. (Eds.). (1994). *Non-Agricultural Employment in India: Trends and Prospects*. New Delhi: Sage Publications.

Vyas, V. S. and Bhargava, P. (1995, October 14–21). Public Intervention for Poverty Alleviation. *Economical and Political Weekly, 30*(41 and 42), 2559–2572.

6 Raising agricultural productivity with reduced use of energy and groundwater

Role of market instruments and technology[1]

M. Dinesh Kumar, Christopher A. Scott and O.P. Singh

Introduction

For countries in the semi-arid and arid tropics, sustainability of agricultural production is largely dependent on irrigation water security (Grey and Sadoff, 2007; Shah and Kumar, 2008). Many such regions are primarily dependent on groundwater for irrigation (Schiffler, 1998). Developing countries, with disproportionate semi-arid and arid tropical regions, face problems of groundwater overdraft due to excessive irrigation withdrawals, which threaten the sustainability of agricultural production and livelihoods of millions of rural families depending on it. India, China, Mexico, Oman, Iran, Pakistan and Morocco are just some among them trying to tackle groundwater overdraft through a variety of approaches (Garduño and Foster, 2010; Giordano and Villholth, 2007; Kumar, 2007; Kumar, Scott and Singh, 2011; Scott, 2011; Scott and Shah, 2004). In such regions, energy and water security are inextricably mixed.

This linkage is patently seen in Indian agriculture. During 2003–04, around 12.8 million electric pumps with a total of 51.84 giga watts (GW) of connected load consumed nearly 87.09 billion kilowatt-hours of electricity (www.indiastat.com). Agriculture accounted for almost 21% of the total power consumption in India. But, in states such as Haryana, Gujarat and Punjab, it was as high as 40.6%, 29.6% and 27.4%, respectively (Zekri, 2008).While in regions with poor electrification but endowed with shallow aquifers, diesel pumps are also put to service for pumping water. As irrigation becomes increasingly energy intensive, energy security is critical for ensuring agricultural water security. Whereas our ability to provide reliable and adequate energy supplies for other sectors of the economy is heavily dependent on how efficiently water for crop production is managed, unmanaged water demand for irrigated agriculture can pose serious challenge to energy security in India (US Energy Information Administration, 2011).

India's farm sector sustains livelihoods for hundreds of millions of rural people; ensures food security for well over a billion; and faces serious management challenges for land, water and energy resources. Carrying on with the marvel of India's rising prosperity rests on getting the groundwater equation right. Raising social and economic equity nationally and minimizing inter-regional disparities will require continued growth of agriculture. At the same time, the farm sector must internalize its share of the effects of groundwater depletion and bankrupt power utilities. Agricultural power – supplied flat rate or free and viewed as an entitlement – must increasingly be managed as a scarce input (World Bank, 2001).

In arid and semi-arid regions of India, groundwater withdrawal for crop production exceeds the average annual recharge. Uncontrolled withdrawal of groundwater for crop production, which is supported by subsidized electricity in the farm sector, leads to rapid declines in water level in many parts of the country (Kumar, 2007; World Bank, 2010). As irrigation is the main user of groundwater in the country, raising water productivity in groundwater-irrigated areas to reduce total water use is essential for arresting groundwater depletion (Amarasinghe et al., 2005; Kumar, 2005, 2007).

Electricity to the farm sector in India is subsidized under both flat rate and pro rata tariff systems (Scott and Sharma, 2009). The subsidy in terms of sale to agricultural consumers has increased from INR 33,363 crore in 2007–08 to INR 45,561 crore in 2011–12 (GOI, 2011).While the extent of subsidy per unit of electricity increased only marginally during the reporting period (from Rs.3.26/kWh to Rs. 3.33/kWh), this enormous increase in total subsidy is due to increasing use of electricity for groundwater pumping,[2] which is the result of an aggregate increase in groundwater draft and increased energy requirement for pumping a unit volume of groundwater. In most states, farmers pay electricity charges based on connected load and not on the basis of units of power consumed. Some of the Indian states are providing electricity to the farm sector free of cost, though with ever-decreasing hours of supply and deteriorating quality of power that results from the poor financial condition of the state electricity boards (SEBs). Modes of electricity pricing, under which the charges paid by farmers do not reflect actual consumption, create incentives for inefficient and unsustainable use of both power and groundwater (Kumar, 2005; Kumar and Singh, 2001).

There have been some developments in metered power supply in the recent past. For instance, nearly 45% of the agricultural power connections are metered in Gujarat now. In West Bengal, the state power board has installed electronic meters in all farm wells and started charging for electricity on the basis of the actual number of units consumed. But empirical studies on the impact of such policy interventions on efficiency, access equity and sustainability in resource use were lacking. Many Indian states are contemplating re-introduction of electricity metering in the farm sector to manage groundwater demand. The basic contention is that at higher power tariff, with induced marginal cost of electricity and water, the farmers will improve water use efficiency (Kumar and Singh, 2001; Kumar, 2005; World Bank, 2001) and enhance water productivity. Such proposals face fierce resistance from farmers' lobby. Further, political parties and scholars alike argue that it will lead to a collapse of farming and the loss of rural livelihoods in many water-scarce regions due to reduced net farm returns, making electricity metering in the farm sector socially and economically unviable.

Energy and carbon footprint of irrigated agriculture in India

India is one of the largest consumers of electricity in the agriculture sector. The largest user of electricity in the agriculture sector is groundwater irrigation, the other being pumping of water from canals and rivers and ponds/tanks. The electricity consumption in agriculture has been steadily going upward due to rapid increase in groundwater abstraction for irrigation and gradual decline in the water table in areas where energized pump sets largely exist. However, the increase has been exponential since 1985–86, and this growth continued till 1998–99 (Figure 6.1).

Since then, it has shown some decline till 2001–02, and then gradually picked up to become 107.77 billion units in 2008–09. But, in percentage terms, the agricultural electricity consumption has begun to decline sharply and consistently since 1998–99 from a highest point of around 31.4% (CSO, 2012). The reason for this is the exponential rise in power consumption in the manufacturing sector, which grew at a rate of 7.4% since 1992–93, coinciding with the year of economic liberalization.

Though declining in percentage terms, agriculture continues to be a major source of India's energy footprint and its contribution in aggregate terms is on the rise, with the total electricity consumption in that sector increasing. This poses a huge environmental challenge. We have estimated the total carbon emission for fossil fuel–based electricity generation to be 28 million metric ton per annum.[3] While efficiency improvements in electric pump sets, which are quite low at present, can reduce this footprint, one reason why this does not happen is that farmers do not pay for consumption of electricity on the basis of actual consumption. But, if the farmers have to pay for electricity on a pro rata basis, then they would have incentive to use both electricity and water efficiently and economically. It is important to note that water use efficiency improvements in irrigation can reduce electricity

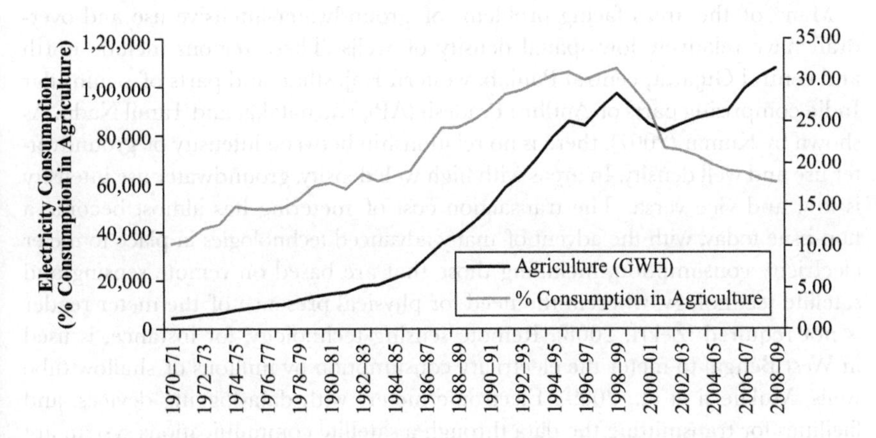

Figure 6.1 Electricity Consumption in Agriculture in India

consumption significantly, whereas pump efficiency improvements through technical interventions will not guarantee water use efficiency improvements, and on the contrary, farmers would be tempted to pump more water. Hence, for reducing carbon emissions, efficient pricing of electricity in the farm sector is important.

Arguments against metering and pro rata pricing

The dominant argument against the shift in power pricing by Shah (1993) is the higher marginal cost of supplying metered electricity owing to the high transaction cost of metering. This may reduce the net social welfare as a result of reductions in: 1) demand for electricity and groundwater in irrigated agriculture and 2) net surpluses that individual farmers generate from farming (Shah, 1993). The second argument, by Saleth (1997), is that for the power tariff to be in the responsive (price-elastic) range of the power demand curve, prices have to be so high that they become socially unviable or politically untenable. The third argument is that under pro rata tariff, the increased cost of pumping groundwater would be transferred to the water buyers (Mukherji et al., 2009). They argue that under flat rate tariff, the water buyers would gain from low irrigation water charges due to competitive water markets, as well owners would have greater incentive to pump more water.

The transaction cost argument is one of the most pervasive arguments in the electricity-groundwater management debate in India. This has mainly stemmed from the sheer number of groundwater abstraction structures in India, estimated to be around 25 million. But while advancing this argument, little attention is paid to the number of wells in regions that, based on water scarcity, actually require co-management of electricity and groundwater. The fact is that the overwhelming numbers of wells in India are in the Indo-Gangetic belt (Scott and Sharma, 2009). This region, especially the eastern Gangetic plain, does not experience serious long-term problems of groundwater overdraft (GOI, 2005; Sharma, 2009).

Many of the areas facing problems of groundwater-intensive use and overdraft have relatively low spatial density of wells. These regions include north and central Gujarat, central Punjab, western Rajasthan and parts of peninsular India comprising parts of Andhra Pradesh (AP), Karnataka, and Tamil Nadu. As shown by Kumar (2007), there is no relationship between intensity of groundwater use and well density. In areas with high well density, groundwater use intensity is low, and vice versa. The transaction cost of metering has almost become a non-issue today, with the advent of many advanced technologies in place to meter electricity consumption, including those that are based on remote sensing and satellite technology, wherein the need for physical presence of the meter reader is not required (Zekri, 2008). Remote sensing technology, for instance, is used in West Bengal to meter the electricity consumption by millions of shallow tube wells (Mukherji et al., 2009). Electronic meters with data-logging devices, and facilities for transmitting the data through a satellite communications system are used on all 11-kilovolt distribution feeders in Andhra Pradesh for energy audit

(Bhatia and Gulati, 2004). As regards the second and third arguments, i.e., socio-economic viability and equity impacts of shift to pro rata tariff, we will examine their validity in the subsequent section.

But political attempts prompt consistently to link power pricing policies to vote banks, arguing that for ruling governments, any decision to raise power tariff in agriculture would be nothing less than a death knell. While one can see the political interests in subsidizing farm power to please the millions of rural voters, the fact, as noted by Howes and Murugai (2003), Kumar and Singh (2001) and Vashishtha (2006a, 2006b), is that it is the large holders constituting a small fraction of the farming community who actually appropriate the majority share of the subsidy benefits under the flat rate system of charging electricity and under free power. Vashishtha (2006a) showed that in AP, large and very large farmers together received 73% of the subsidy benefits in 2003–04, whereas small and marginal farmers received only 5.1%. The corresponding figures for Punjab were 73.8% and 6.3%, respectively (Vashishtha, 2006b). While a large majority of the farming community does not gain from such policies, the politicians are largely ignorant about this. Understanding the nuances of this would go a long way in convincing the political leaders about the need to do away with such policies that have negative effects on productivity and equity – i.e., they neither benefit the large section of the rural masses nor help improve the water and energy economies of the states concerned.

Some scholars cite positive impact of flat rate pricing of electricity on access and equity of groundwater (for instance, Shah, 1993). They argue that with competitive water markets that emerge as a result of flat rate pricing, water prices would be low with the result that a major share of the electricity subsidy benefits are transferred to water buyers. However, the zero (or extremely low) marginal cost of production of water from wells does not seem to influence the prices at which water is traded, in favor of buyers of water for irrigation. Recent research shows that flat rate pricing increases the monopoly power of large well owners (Kumar, Singh and Singh, 2001). Flat rate pricing also leads to inequitable distribution of power subsidy benefits among well owners (Kumar and Singh, 2001; Howes and Murugai, 2003). Kumar (2007), on the basis of evidence from Muzaffarpur in eastern Bihar argued that the monopoly power enjoyed by water sellers cannot be reduced by pricing policies, but by improving the transferability of groundwater.

In an attempt to cope with the increasing financial burden due to revenue losses through subsidies and growing power deficits, the state electricity boards in many agriculturally prosperous states have introduced heavy cuts in power supply hours to the farm sector (GOI, 2002). Examples are Punjab, Andhra Pradesh and Gujarat. The assumption here is that this would reduce the energy use and groundwater draft for agriculture. The electricity boards have not analyzed the impact of such cuts on equity in access and efficiency in use of groundwater. On the contrary, with reduction in hours of power supply, the quality of irrigation can be adversely affected.[4] The economic prospects of irrigated farming are more elastic to the quality of irrigation water than to its cost (Kumar and Patel, 1995; Kumar and Singh, 2001). The rich well owners always find ways to overcome the crisis of power cuts. This can further increase their monopoly in water trading.

Research objectives, study location, approach and methodology

The broad objective of the research here is to analyze the socio-economic viability of pro rata pricing of electricity in agriculture and to assess various technological options for implementing energy pricing policies. Specific objectives are: 1) to study the impact of the shift from flat rate power supply to metered supply on the efficiency and sustainability of groundwater use by well owners; 2) to analyze the overall impact of electricity pricing on the farming system of well owners, including the economic returns from farming; and 3) to discuss various alternatives for implementing energy pricing policies and their likely outcomes vis-à-vis sustainability and efficiency of groundwater use, and equity in access to groundwater.

North Gujarat, which is a water-scarce region, and the eastern plain regions of Uttar Pradesh (UP) and south Bihar, which are water-rich regions, are the study locations. Water-rich regions of UP and Bihar were selected for the study due to the reason that there were no other locations in India where a comparison could be made between farmers who are confronted with marginal cost of using energy and groundwater for irrigation, and farmers who are not confronted with this problem. The semi-arid north Gujarat region receives a mean annual rainfall of 735 mm. Grey-brown, coastal alluvium types of soils are found in this region. The mean annual precipitation in the eastern plain region of UP is about 1025 mm and the region's climate varies from dry sub-humid to moist sub-humid. The soil type in this sub-zone is light alluvial and calcareous clay. South Bihar plains receive a mean annual rainfall of 1103 mm and the climate condition of the region varies from dry to moist sub-humid. The soil types found in the region are old alluvium sandy loam to clayey, and the larger areas under traditional water storage and irrigation systems are called *Tal* and *Diara*.

Primary and secondary data relating to crop and livestock production were obtained through surveys. The primary data included extent of crop inputs and outputs and their prices, cropping pattern, electricity prices, diesel consumption and price, well command area, number of water buyers and sellers, quantum of livestock inputs and outputs and unit price of inputs and outputs. Banaskantha district in north Gujarat, Mirzapur and Varanasi districts in eastern UP, and Patna district in south Bihar were selected for the study. The details of the sample design for each location are given in Table 6.1. At the time of undertaking this study, there were very few locations in India where farmers paid for electricity based on consumption. Gujarat was one such state. Therefore, to analyze the potential impacts of introducing pro rata pricing of electricity in farm sector in the other states, farmers using diesel pumps for groundwater irrigation and water buyers were selected as a proxy for pro rata tariff.

Table 6.1 Sampling Procedure and Sample Size

Name of the Region	*Name of the District*	*Type of Energy Tariff*				*Diesel Pump*		*Total Sample Size*
		Flat Rate		*Pro Rata*		*Well Owners*	*Water Buyers*	
		Well Owners	*Water Buyers*	*Well Owners*	*Water Buyers*			
North Gujarat	Banaskantha	60	–	60	–	–	–	120
Eastern UP	Varanasi and Mirzapur	60	60	–	–	60	60	240
South Bihar	Patna	60	60	–	–	60	60	240
Total		**180**	**120**	**60**	**–**	**120**	**120**	**600**

Source: Primary survey undertaken in the selected regions

The price of electricity used for pumping groundwater influences water productivity in many different ways. They are equity in access, efficiency of use of water, economic viability of farming and sustainability of groundwater use (Kumar, 2005). The efficiency impact of change in mode of pricing was analyzed by comparing water productivity of crops in physical terms. The impact of change in mode of pricing on economic viability of farming was analyzed by comparing the overall water productivity of crops, livestock and farming system in economic terms under the two conditions. The net return from unit area of land farmed was also considered. The net return was based on the cost A_2. The net income was estimated by subtracting 'Cost A_2' from gross income from the crop (crop produce in kg × price received by the farmers/kg). Cost A_2 is Cost A_1+ rent paid for leased in land. Here, Cost A_1= wages of hired, contract and permanent labor + hired bullock labor/imputed value of own bullock labor + charges of hired machinery/imputed value of owned machinery + market rate of organic manure and fertilizers + market rate of seed/imputed value of owned seed + imputed value of manure + market value of insecticide, herbicide + irrigation charges + land revenue, cess and other taxes + depreciation of machinery, implements, equipment, irrigation structure + interest on working capital + miscellaneous expenses.

The sustainability impact of price changes is analyzed by looking at the changes in groundwater withdrawal for unit irrigated area by well-owning farmers. Here, we have considered the applied (pumped) water for estimation of water productivity at the field- and farming-system level, and not the depleted water that takes into account the contribution of rainfall to total water input to the crop and return flows into groundwater. This does not alter the inferences drawn from the study due to three reasons. First, we are concerned with the changes in water productivity in the same field or farm, which means that the level of use of rainfall by the crop does not change. Second, if rainfall use increases, it will not change the groundwater recharge component of irrigation. Third, return flows would be insignificant in semi-arid north Gujarat due to deep water table conditions (Kumar et al., 2008a). Though return flows can be quite significant in both UP and Bihar plains due to alluvial formations and sub-humid climatic conditions, the farmers in these regions would be concerned with the total amount of water applied rather than the actual amount of water depleted. The reason is that applied water would determine the amount of energy required to pump groundwater, which is scarce in these regions.

The physical water productivity for a given crop (kg/m^3) is estimated using data on crop yield and the estimated volume of water applied for all sample farmers growing that crop. The volume of water applied to the crop was estimated from the discharge of the wells owned by the farmers (including those who sell the water) (Q); number of irrigations given to the crop (n); and duration of watering per irrigation (t) as $Q \times n \times t$. The discharge of the wells was measured in the field using a stopwatch and a bucket with known capacity, by allowing the output of the well to fill directly in the bucket and then noting the time required to fill the bucket. The combined physical and economic water productivity in Rs/m^3 is estimated using data on net returns from crop production in Rs/ha and estimated volume of water. To estimate the net income from a particular crop, the data on inputs for each crop were obtained by primary survey of farmers. These included cost of seed, labor, fertilizer, pesticides and insecticides, irrigation, plowing, harvesting and threshing.

The physical productivity of water in milk production for livestock is estimated using the methodology presented in Kumar (2007) and Singh (2004). The water productivity in farming operations, including crops and dairying, is estimated using the methodology presented in Kumar et al. (2008b), which was used for estimating the economic value of irrigation water in agriculture for individual farms.

Results and discussion

Distribution of land holdings

In north Gujarat, the average size of land holdings is higher for tube well owners who are paying power tariff on connected load basis (3.45 ha) as compared to their counterparts with metered connections (2.95 ha). About 90% of the area is under irrigated crop production and the remaining 10% area is cultivated under rain-fed conditions.

In eastern UP, the average size of land holdings is larger for diesel well-irrigated farms[5] as compared to electric well commands. Differences are significant between well owners and water buyers. Diesel pump owners have an average land holding size of 1.35 ha, while their water buyers have a landholding size of 0.94 ha. The average size of land holdings for electric pump owners is 1.30 ha, whereas their water buyers have an average land holding size of 0.56 ha.

In south Bihar, the average size of land holdings for both well owners and water buyers in the diesel pump commands are higher than that of their electric counterparts. The well owners in electric well-irrigated farms have larger sized holdings (0.73 ha) as compared to their water buyers (0.53 ha). In diesel pump commands, the differences are larger. The average size of land holding of well owners here is 1.26 ha, whereas for water buyers it is 0.57 ha.

Hence, the average size of land holding in water-rich eastern UP and south Bihar plains is much smaller when compared to water-scarce north Gujarat. This is one of the important factors that determine the utilization of available water resources. In case of water-abundant regions, the limited land availability should motivate farmers to maximize returns per unit of land. Against this, in water-scarce regions, water availability is a limiting factor for maximizing returns from crop production, and hence generally, they would be motivated to maximize the returns from every unit of water (Kumar et al., 2008b). However, lack of resources for investing in wells and energizing devices is a limiting factor for many farmers in south Bihar and eastern UP to access the water.

Cost of groundwater pumping

The cost of groundwater pumping for well owners was estimated by taking into account the following: 1) cost of well construction and pump set installation, 2) cost of obtaining power connection, 3) cost of operation and maintenance of the well and the pump set, 4) life of the well and the pump set, 5) the average hours of groundwater pumping per year and 6) discharge of the pump set. Since the year of construction is not the same for all wells, the cost of construction of each well was adjusted to inflation to make it correspond to the prices in the base year, i.e., the year of study (2008), and then discounted for the life of the system (20 years) to get annualized costs, using discount rate. To this, the variable cost (cost of operation and maintenance) was added to obtain the annual cost of irrigation. This, when divided by the annual hours of irrigation and the pump discharge (m^3/hour), yields the cost of groundwater irrigation per m^3 of water.

In the case of electric wells with metered connections, the operational cost (per hour) is worked out using the energy charges per kilowatt-hour (kWh) of use. Similarly, in the case of diesel wells, the operation cost was worked out using the price of one liter of diesel and the amount of diesel consumption per hour of running. The cost of irrigation per cubic meter of water was finally worked out using well output data. In the case of wells with flat rate electricity connection, the implicit cost per hour of irrigation is worked out using the annualized cost, and the number of hours of irrigation per annum. Based on the figures of well discharge, cost estimates were worked out for eastern UP, northern Gujarat and south Bihar and are presented in Table 6.2. The unit rates charged by diesel pump owners for irrigation services are much higher than those of electric pump owners.

Table 6.2 Cost of Irrigation Water for Different Categories of Farmers From the Three Study Locations

Area	*Water Source*	*Average (Rs/m^3)*	*Range (Rs/m^3)*
Eastern UP	Electric pump owner	0.18	0.10–0.30
	Electric pump buyers	0.65	0.52–0.84
	Diesel pump owners	1.38	0.99–2.04
	Diesel pump water buyers	2.81	2.07–3.63
North Gujarat	Metered connections	1.07	0.14–3.91
	Nonmetered connections	1.60	0.19–4.27
South Bihar	Electric pump owner	0.77	0.17–3.39
	Electric pump water buyers	0.70	0.31–0.92
	Diesel pump owners	1.87	1.51–2.95
	Diesel pump water buyers	2.15	1.84–2.42

Source: Calculated from authors' primary data

Cropping patterns

Table 6.3 shows the cropping pattern of well owners and water buyers under different modes of energy pricing, i.e., connected load (electric well) and unit consumption (diesel well) in eastern UP. Paddy and wheat are the dominant crops. The crops grown in the study villages are food grains, pulses, oilseeds, vegetables, cash crops and fodder crops. During the kharif season, diesel and electric well owners and water buyers allocate larger portions of their land holding under paddy.

In diesel well commands, pump owners allocate about 26% of the gross cropped area to paddy cultivation, whereas buyers of water from diesel well owners allocate only 22%. In electric well commands, pump owners allocate 12% to paddy, and water buyers allocate about 15% to paddy. Electric pump owners also grow groundnut. Water buyers in both electric and diesel well commands allocate larger portions of their cropped area under green fodder and other vegetables during kharif season as compared to pump owners. Water buyers in diesel well commands grow lentils of the *Arhar* variety. Water buyers in electric well commands grow lady's finger (okra).

Major crops grown during winter season are wheat and barley, potato, pea, gram, mustard, linseed and barseem. The percentage area allocated for crops, viz., wheat, pea, potato and barseem, is lower for well owners as compared to water buyers, whereas well owners allocate larger area to crops, viz., mustard, gram, barley and linseed, as compared to water buyers.

In diesel well commands, pump owners allocate larger shares of their cropped area under winter crops as compared to water buyers. Such sharp differences are not seen in the case of electric well commands. During the summer season, major

Table 6.3 Cropping Patterns of Well Owners and Water Buyers Under Different Energy Regimes, Eastern UP

Name of the Crops	*Electric Well Command*				*Diesel Well Command*			
	Owner		*Water Buyers*		*Owner*		*Water Buyers*	
	Area (Ha)	*% Area*	*Area (Ha)*	*% Area*	*Area (Ha)*	*% Area*	*Area (Ha)*	*% Area*
Kharif Season								
1. Paddy	0.71	11.51	0.36	14.81	1.55	26.18	0.91	22.14
2. Bajra	0.32	5.15	0.14	5.85	0.23	3.85	0.13	3.25
3. Maize	0.24	3.97	0.12	4.78	0.23	3.81	–	–
4. Lady's finger	0.32	5.18	0.23	9.53	–		–	
5. Other vegetables	0.32	5.30	0.17	7.08	0.14	2.41	0.34	8.35
6. Arhar	–		–		–	–	0.30	7.42
7. Black gram	0.27	4.39	0.11	4.68	–	–	0.11	2.78
8. Green gram	0.37	6.06	–		–	–	0.11	2.78
9. Sesame	0.08	1.30	0.06	2.34	0.23	3.85	0.11	2.78
10. Groundnut	0.33	5.34	–		–	–	–	–
11. Sugarcane	0.11	1.77	0.06	2.34	0.16	2.68	–	–
12. Green fodder	0.16	2.60	0.08	3.20	0.11	1.89	0.10	2.38
Rabi Season								
1. Wheat	0.67	10.94	0.29	12.00	1.27	21.48	0.83	20.29
2. Barley	0.23	3.73	0.08	3.28	–	–	0.09	2.23
3. Pea	0.23	3.80	0.13	5.47	0.34	5.73	0.17	4.08
4. Gram	0.17	2.85	0.04	1.46	0.42	7.02	0.20	4.84
5. Mustard	0.70	10.06	0.53	4.45	0.27	4.55	0.14	3.50
6. Linseed	0.06	0.93	–	–	0.34	5.78	0.10	2.50
7. Potato	0.50	8.15	0.29	11.94	0.37	6.24	0.23	5.57
8. Barseem (Green fodder)	0.07	1.14	0.05	1.89	0.06	1.05	0.07	1.64
Summer Season								
1. Sunflower	0.10	1.58	–	–	–	–	–	–
2. Vegetables	0.11	1.86	–	–	0.11	1.93	–	–
3. Green fodder	0.15	2.38	0.12	4.89	0.09	1.55	0.14	3.48
Gross cropped area (GCA)	**6.13**	**100.00**	**2.44**	**100.0**	**5.92**	**100.00**	**4.10**	**100.0**

Source: Calculated from authors' primary data

crops grown in electric well commands are green fodder, sunflower and vegetables. While all these crops are grown by the electric pump owners, water buyers grow only green fodder. In diesel well commands, crops grown during summer season are green fodder and vegetables. Both diesel well owners and water buyers are found to be growing green fodder in some portion of their land.

In the case of north Gujarat, major crops grown by the tube well owners under both tariff regimes are green fodder, food grain crops, pulses, groundnut and cash crops, such as cluster bean, cotton and castor. The farmers of this region allocate small area for green fodder throughout the year.

During kharif, tube well owners under metered tariff regime allocate a slightly larger percentage of the cropped area to cotton, castor and fodder bajra. During winter, tube well owners under flat rate tariff regime are allocating more area under green fodder, wheat and mustard. The tube well owners under pro rata tariff regime allocate slightly larger area under cumin, which is a high-valued cash crop. The major crops grown during summer season are green fodder and bajra. The tube well owners under both flat rate and unit tariff regime have about 10% of their gross cropped area under bajra.

In south Bihar, which receives very heavy monsoon rains, during kharif, farmers grow paddy and green fodder, with larger area under paddy. Out of the gross cropped area, nearly 38% is under paddy. During winter, farmers grow wheat, gram, mustard, barseem (fodder), potato, radish, carrot and coriander. During summer, they grow onion, maize and green fodder. There is no significant difference in kharif cropping pattern between well owners and water buyers in electric well commands or diesel well commands. During winter, water buyers in electric well commands cultivate gram and carrot. Diesel pump owners and water buyers in both diesel and electric well commands keep a larger area for growing potato. During summer, only diesel pump owners and water buyers in their commands cultivate green fodder. In general, electric pump owners allocate larger area under different crops as compared to their water-buyer counterparts. There is a similar trend in the case of diesel pump command areas.

Irrigation and crop water productivity

Higher physical productivity of water use for a given crop indicates more efficient use of irrigation water through on-farm water management or better farm management. Higher water productivity in economic terms means better economic viability of irrigated production, if surplus land is available for expanding irrigated area.

Figure 6.2 presents the estimates of irrigation water dosage (cm) for winter crops of pump owners and water buyers in electric well commands in eastern UP villages. Figure 6.3 presents the estimates of water productivity of crops in physical (kg/m^3) and economic terms (Rs/m^3) for pump owners and water buyers in these commands. As Figure 6.2 shows, the total amount of irrigation water applied for crop production is higher for electric pump owners as compared to irrigation water buyers. Further, as Figure 6.3 shows, for most of the crops, both physical and economic productivity of water are higher for water buyers than well owners.

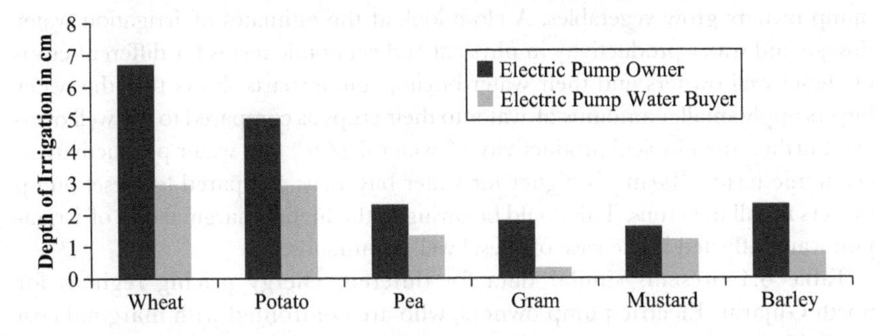

Figure 6.2 Depth of Irrigation in Rabi Season (Eastern UP) for Pump Owners and Water Buyers in Electric Well Command

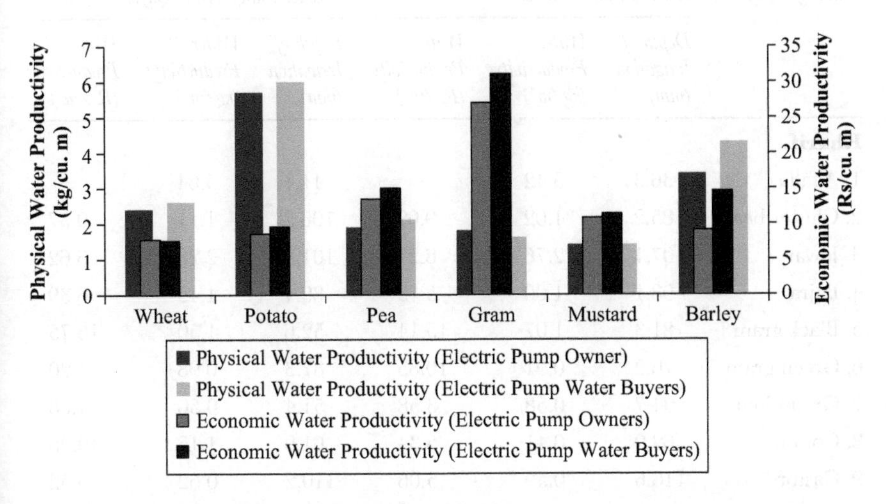

Figure 6.3 Water Productivity in Rabi Season (Eastern UP) for Pump Owners and Water Buyers in Electric Well Command

The fact that water buyers do not grow crops during summer when crop water requirement generally becomes high is equally important. During this season, well owners grow water-intensive vegetable crops.

As regards diesel well commands, though both the well owners and water buyers are confronted with marginal cost of using water, the water buyers incur higher cost. But there is not much difference in the cropping pattern of pump owners and water buyers, except that water buyers do not grow sugarcane and maize. To economize on irrigation water, water buyers cultivate water-efficient crops such as arhar, black gram and green gram during kharif season. The winter cropping pattern is the same for diesel pump owners and water buyers. During summer, only pump owners grow vegetables. A close look at the estimates of irrigation water dosage and water productivity in physical and economic terms for different crops of diesel well owners and their water-buying counterparts shows that the water buyers apply smaller amounts of water to their crops as compared to the well owners. Further, the physical productivity of water (kg/m^3) and water productivity in economic terms (Rs/m^3) is higher for water buyers as compared to diesel pump owners for all the crops. This could be owing to the higher marginal cost of irrigation water affected in the case of diesel well commands.

Table 6.4 presents similar data for different energy pricing regimes for north Gujarat. Electric pump owners, who are confronted with marginal cost

Table 6.4 Water Use, and Water Productivity in Physical and Economic Terms Under Flat and Unit Energy Pricing Regimes, North Gujarat

Name of Crop	*Electric Pump Owner*			*Electric Pump Water Buyer*		
	Depth of Irrigation (mm)	*Water Productivity (kg/m^3)*	*Water Productivity (Rs/m^3)*	*Depth of Irrigation (mm)*	*Water Productivity (kg/m^3)*	*Water Productivity (Rs/m^3)*
Kharif						
1. Alfalfa	36.3	5.42	–	41.1	5.64	–
2. Cluster bean	85.2	1.02	9.09	106.2	1.11	9.37
3. Jowar	107.1	2.76	8.27	101.4	2.26	6.62
4. Bajra	98.1	1.00	5.13	89.4	1.45	6.39
5. Black gram	81.3	1.07	15.14	52.6	1.50	16.75
6. Green gram	76.2	0.91	10.85	87.3	0.98	11.20
7. Groundnut	94.7	0.58	3.58	51.4	0.56	4.68
8. Cotton	62.9	0.41	5.34	61.0	1.15	19.28
9. Castor	116.6	0.59	5.06	110.2	0.62	6.52
Rabi						
1. Alfalfa	32.7	3.65	–	28.3	5.71	–
2. Wheat	127.2	0.82	4.64	96.3	0.91	5.17
3. Barley	22.9	0.47	0.70	62.9	1.11	6.17
4. Rajgaro	91.4	0.56	4.11	72.7	0.89	8.50
5. Mustard	113.8	2.86	22.25	74.6	2.10	23.50
6. Cumin	89.5	0.82	36.71	81.4	0.99	47.71
Summer						
1. Alfalfa	38.2	2.30	–	–	–	–
2. Bajra	168.7	1.95	6.43	129.2	1.94	7.31

Source: Calculations from authors' primary data

of using electricity, maintain higher water productivity in both physical and economic terms for all the crops as compared to those who are paying for electricity on the basis of connected load (pump horsepower). Further, they do not grow highly water-intensive alfalfa, which is a fodder, in their fields during summer.

Comparison of estimates of mean values of irrigation water dosage and water productivity in physical and economic terms for crops of both pump owners and water buyers in electric pump commands in south Bihar plains (Table 6.5) shows that water buyers apply less water to their crops and maintain higher physical water productivity for many crops in comparison to electric well owners. However, they secure lower water productivity in economic terms for most of the crops, except radish and onion. This could be due to the higher cost of irrigation water, which eventually reduces net return from crop production.

Comparing the water use and water productivity of crops raised by pump owners and water buyers in diesel well commands of south Bihar plains – both in physical and economic terms – shows that both categories grow almost similar crops. For all crops except onion and summer green fodder, water buyers in diesel

Table 6.5 Water Use, and Water Productivity in Physical and Economic Terms Under Electric Well Command, South Bihar Plains

Name of Crop	*Electric Pump Owner*			*Electric Pump Water Buyer*		
	Depth of Irrigation (mm)	*Water Productivity (kg/m³)*	*Water Productivity (Rs/m³)*	*Depth of Irrigation (mm)*	*Water Productivity (kg/m³)*	*Water Productivity (Rs/m³)*
Kharif						
1. Paddy	75.1	2.5	6.35	46.7	2.69	8.4
2. Maize	25.0	20.5	–	12.5	27.34	–
Rabi						
1. Wheat	48.2	1.8	5.56	35.1	1.76	5.8
2. Potato	192	13.1	43.16	20.0	11.74	41.8
3. Barseem	5.6	10.4	–	4.0	11.91	–
4. Mustard	26.7	1.8	20.16	–	–	–
5. Gram	–	–	–	9.3	0.66	9.2
6. Radish	12.7	10.0	13.92	9.6	9.59	18.5
Summer						
1. Onion	46.0	4.4	18.48	21.8	5.40	23.2
2. Maize	20.7	5.9	21.66	17.6	6.86	19.1

Source: Calculated from authors' primary data

well commands secure higher physical water productivity as compared to pump owners. Again, for all crops except onion, the water buyers secure higher water productivity in economic terms as compared to pump owners.

The trends emerging from the foregoing analysis are as follows: 1) water productivity in economic terms of water buyers in electric pump commands is greater than that of well owners both in eastern UP and south Bihar; 2) physical productivity of water for electric pump owners under flat rate tariff is comparatively less than that under pro rata tariff; 3) water productivity in economic terms of electric pump owners is less than that of diesel pump owners; and 4) economic productivity of water for water buyers of electric pumps is less than that of those buying water from diesel well owners.

Water productivity in dairying

Estimates of weighted average of feed and fodder input to livestock for the entire animal lifecycle by farmers were carried out for farmers in electric and diesel well commands in eastern UP, farmers with metered and non metered power connections in north Gujarat, and farmers in the electric well commands and diesel well commands in south Bihar. The overall observation is that water buyers in eastern UP and south Bihar and farmers in north Gujarat with non metered connections fed more input to their cattle.

The figures of average milk production from dairy animals worked out for the entire animal life cycle for electric well owners of eastern UP are 2.91, 4.64 and 1.81 liters/day/animal for buffalo, crossbred cow and indigenous cow, respectively. The corresponding estimates for farmers in the diesel well commands are 2.08, 4.01 and 1.95 liters for well owners; and 2.23, 3.23 and 2.01 for water buyers, for buffalo, crossbred cow and indigenous cow, respectively.

In north Gujarat, in the case of farmers having metered electricity connections, the average milk production from buffalo, crossbred cow and indigenous cow are 5.14, 7.50 and 1.91 liters/day/animal, respectively. Similar estimates for non metered connections are higher at 6.96, 9.32 and 6.43 liters. Such higher yields in the case of dairy animals owned by farmers with flat rate connections are because of the higher amount of feed and fodder that they are being provided with. Since they are not confronted with marginal cost of using electricity and groundwater, they grow water-intensive alfalfa, which has high nutrition value, more extensively, and feed their animals.

In the case of electric pump owners in south Bihar, the average milk production figures from buffalo, crossbred cow and indigenous cow are 2.0, 2.36 and 0.79 liters/day/animal, respectively. In the case of water buyers, they are 1.86, 2.97 and 0.88 liters/day/animal. The figures for farmers of diesel well owners are 1.69, 3.53 and 0.96 liters/day/animal, respectively, whereas, in case of water buyers, the corresponding values are 1.68, 2.30 and 1.18 liters/day/animal.

The estimates of the volume of water used for milk production, physical productivity of water in milk production and gross water productivity in milk production in economic terms for buffalo, crossbred and indigenous cows for the sample

Table 6.6 Water Use, Input Costs and Water Productivity (Physical and Economic) in Milk Production in Electric Pump Command Area, Eastern UP (m^3/day)

Types of Feed and Fodder	*Electric Pump Owner*			*Electric Pump Water Buyer*		
	Buffalo	*Crossbred Cow*	*Indigenous Cow*	*Buffalo*	*Crossbred Cow*	*Indigenous Cow*
1. Total water used for feed, fodder and drinking (m^3)	2.63	3.02	2.50	2.19	3.35	2.38
2. Milk production (Lt)	2.91	4.64	1.81	2.64	4.08	1.89
3. Milk WP (Lt/m^3)	1.11	1.54	0.72	1.20	1.22	0.79
4. Gross WP (Rs/m^3)	11.95	15.52	6.72	12.97	12.31	7.35
5. Total expenditure (Rs/day)	12.84	15.52	12.84	12.73	19.80	14.03
6. Milk production (Lt)	2.91	4.64	1.81	2.64	4.50	1.89
7. Gross income – milk and dung (Rs)	31.89	47.36	17.33	28.94	45.98	18.01
8. Net income (Rs/day)	19.05	31.84	4.50	16.21	26.18	3.98
9. Net water productivity (Rs/m^3)	7.25	10.55	1.80	7.39	7.82	1.67

Source: Calculated from authors' primary data

farmers in the electric well commands in eastern UP are presented in Table 6.6. The estimates of total water used for animals consider the embedded water in feed and fodder. Detailed discussion on embedded water in milk production can be seen in Singh (2004) and Kumar (2007). Dairy farmers, who own pumpsets, use larger quantities of water for producing green and dry fodder, in comparison to water buyers. However, the amount of water embedded in the concentrate used for dairy production is higher for water buyers. The net result is that the gross water productivity for milk production is higher for electric pump owner as compared to water buyers.

The volume of water used for milk production by water sellers in diesel well command are 3.02 m^3, 3.48 m^3 and 2.68 m^3/day/animal for buffalo, crossbred cow and indigenous cow, respectively. The corresponding figures for water buyers are 3.00 m^3, 3.21 m^3 and 2.64 m^3/day/animal. The physical productivity of water for milk production are 0.69, 1.15 and 0.73 liter/m^3, respectively, for pump owners and 0.75, 1.00 and 0.76 liter/m^3 for water buyers. Similar estimates are available for the two categories of farmers in north Gujarat, and four categories of farmers in south Bihar.

The net water productivity in economic terms for dairy production was estimated by considering the cost of milk production, which includes the cost of production of dry fodder, green fodder, cattle feed and other expenses for maintaining dairy animals in the water productivity analysis. Based on these data, the net water productivity in economic terms was estimated for all the three locations, i.e., for well owners, and water buyers in electric and diesel well commands in eastern UP and south Bihar, and electric well owners with and without metered connections in north Gujarat.

The results for farmers in electric commands in eastern UP (Table 6.6) show that across livestock types, well owners secure higher net water productivity in milk production than water buyers. The values of net water productivity in economic terms for the diesel well owners are 1.74, 6.89 and 0.46 for buffalo, cross bred cow and indigenous cows, respectively. The corresponding values for water buyers are 0.43, 1.8 and −1.72. Comparing electric and diesel well commands, it appears that electric well owners secure highest net water productivity in economic terms, followed by water buyers in their command, diesel pump owners and lowest for buyers of water from diesel pump owners.

In north Gujarat, the average values of net water productivity in economic terms for milk production from buffalo, crossbred cow and indigenous cow for farmers under flat energy pricing regime are Rs 3.73/m^3, Rs 5.88/m^3 and Rs −1.85/m^3, respectively. Against these, the values for farmers under pro rata pricing regime are Rs 3.31/m^3, Rs 2.29/m^3 and Rs 3.37/m^3, respectively. Thus, overall net water productivity is higher under a pro rata pricing regime, wherein the farmers have to pay for every unit of water used for growing fodder crops and cereals indirectly through electricity charges.

In south Bihar, the estimates of average water productivity in economic terms in dairy production for electric well owners are Rs 2.18/m^3, 1.96/m^3 and −1.0/m^3. The corresponding values for water buyers are Rs 1.65/m^3, 3.89/m^3 and −0.64/m^3 for buffalo, crossbred cow and indigenous cow, respectively. For pump owners in diesel well commands, net water productivity in economic terms (Rs/m^3) are −0.47, 5.68 and −2.50; and for water buyers, the values are 0.07, 2.09 and −1.26, for buffalo, cross bred cow and indigenous cow, respectively.

Farm-level water productivity

In eastern UP and south Bihar, farmers should try and economize on the use of water, although it is not a scarce resource in these regions in physical terms. The reason is that using more water means paying more for pump rental services. Farms are the unit for many investment decisions by farmers in agriculture including water allocation decisions. Hence, they try to optimize water allocation over the entire farm, rather than individual crops, to maximize their returns. Therefore, the impact of power pricing on the efficiency with which water is used by farmers should be analyzed by looking at the water productivity for the entire farming system.

The analysis clearly shows that the overall farm-level water productivity is much higher for water buyers in diesel well commands in eastern UP and south Bihar (Figure 6.4). In electric well commands also, the differences exist in favor of water buyers in spite of very low marginal cost of using water

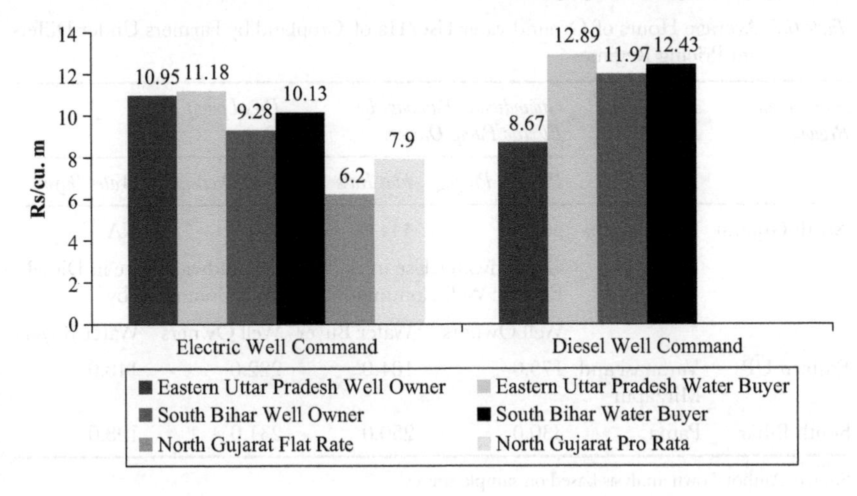

Figure 6.4 Water Productivity in Farming Operations Under Different Pricing Regimes

(Rs 0.65/m^3). The farm-level water productivity is much higher for farmers who are confronted with marginal cost of electricity in north Gujarat as compared to those who pay for electricity based on connected load. The water productivity improvement in highest in eastern UP in the diesel well commands, where the water buyers' marginal cost of using irrigation service is Rs 2.81/m^3. In north Gujarat, the difference in water productivity between farmers with flat rate connection and those with metered connections is quite substantial.

Further, comparison between electric well owners and diesel well owners in both the locations substantiates the earlier point that positive marginal cost promotes efficient use of water at the farm level.

Groundwater pumping and net farm returns of farmers

Pricing would induce efficiency, but may not ensure sustainability of resource use (Kumar, 2005). The total amount of groundwater pumpage per unit of cultivated area is determined by the cropping pattern, the cropping intensity and the degree to which crop water needs are met. Increased allocation of cultivable area under highly water-intensive crops would increase the demand for irrigation water by a farmer. Hence, total pumpage per unit cultivated area could be a good indicator of the sustainability impacts of change in mode of pricing on groundwater. However, farmers with very small land holding size are more likely to intensify cropping, which would increase the total pumpage. This would mean longer hours of pumpage per ha of cultivable area, as the value of numerator would increase and that of denominator would decrease.

The pumpage of groundwater per unit area of cultivated land is found to be lower for water buyers of eastern UP and south Bihar, in spite of their having

Table 6.7 Average Hours of Groundwater Use/Ha of Cropland by Farmers Under Different Pricing Regimes

Name of the Regions	*Name of the District*	*Groundwater Pumpage by Electric Pump Owners*		*Diesel pump*	
		Pro rata Pricing	*Flat Rate*	*Well Owners*	*Water Buyers*
North Gujarat	Banaskantha	304.0	444.00	NA	NA
		Groundwater Use in Electric Well Command by		Groundwater Use in Diesel Well Command by	
		Well Owners	Water Buyer	Well Owners	Water Buyers
Eastern UP	Varanasi and Mirzapur	175.0	184.0	222.0	148.0
South Bihar	Patna	330.0	250.0	231.0	198.0

Source: Authors' own analysis based on sample survey

Table 6.8 Net Income From Farming Operations in the Three Study Locations

Region	*Type of Well Command*	*Type of Farmer*	*Net Income From Crops (Rs)*	*Net Income From Dairying (Rs/day)*	*Total Farm-Level Income (Rs)*	*Farm-Level Income per Unit Land (Rs/Ha)*
Eastern UP	Electric well	Well owner	124,587	7,152	131,740	24,880
		Water buyer	54,638	6,165	60,803	27,570
	Diesel well	Well owner	74,765	7,430	82,194	14,528
		Water buyer	62,323	6,261	68,584	18,075
North Gujarat	Electric well	Flat rate pricing	369,120	30,048	768,287	57,531
		Pro rata pricing	311,807	45,636	669,250	56,882
South Bihar	Electric well	Well owner	120,477	10,293	130,770	210,345
		Water buyer	61,518	8,131	76,024	190,031
	Diesel well	Well owner	140,105	9,958	150,064	191,387
		Water buyer	71,810	12,232	84,043	197,895

Source: Calculated from authors' primary data

lower sized holdings (Table 6.7). The data for north Gujarat show that the pump owners having metered connections, in spite of having smaller sized land holdings (2.95 ha against 3.45 ha) use much less water per ha of land as compared to their flat rate counterparts (304.0 hours per year against 444.0 hours per year). The difference in aggregate pumping is much greater between farmers with meters and those without meters (Table 6.7). Such a high reduction is water usage per unit of cultivated land, which is disproportionately higher than the reduction in net return per unit of land, is made possible through high improvements in water productivity in economic terms.

In spite of a slight reduction in pumping, the net return from unit area of land is higher for water buyers in eastern UP and south Bihar plains (Table 6.8). This is achieved through high improvement in water productivity through selection of crops that are less water consuming and high valued. As Reddy (2009) notes, improving productivity of land and water are the two ways of making pricing affordable to the farmers. Although the net returns per unit of land were marginally lower for farmers who paid on pro rata basis in north Gujarat (Table 6.8), this is not a concern, as in water-scarce regions like north Gujarat farmers would not have land constraints in maximizing returns. Even if the farmers attempt to expand the area to maintain the net farm return at the previous levels, the aggregate water usage would still be lower than the previous levels.

Energy pricing in the farm sector and its implications for groundwater sustainability and power sector viability

As our analysis showed, introducing marginal cost for electricity motivates farmers to use water more efficiently at the field level from physical, agronomic and economic points of view through careful use of irrigation water, use of better agronomic inputs and optimizing costly inputs. This is evident from: 1) the lower irrigation dosage applied by farmers who are either using diesel wells or buying water from well owners or paying for electricity on pro rata basis, with lowest dosage found in the case of water buyers of diesel well commands, who pay higher unit price for irrigation water; and 2) the higher physical and economic productivity of water in crop production secured by the farmers who are either using diesel wells or buying water from well owners, or paying for electricity on pro rata basis.

Further, introducing marginal cost for electricity motivates farmers to use water more efficiently at the farm level through careful selection of low water-intensive crops, and livestock composition that give higher return from every unit of water. This is evident from the higher water productivity secured in farming operations by those who purchase water as compared to their well-owning counterparts, and those farmers who pay for electricity on pro rata basis. The results also showed that higher cost of irrigation water affected by higher energy cost will not lead to lower net return from every unit of water used, as the farmers modify the farming system itself in response to increase in energy cost.

As the analysis suggests, pro rata pricing has significant impact in reducing groundwater pumpage from every unit of irrigated land, which is disproportionately higher than the reduction in net return from unit of land. This means that even if farmers expand the area to maintain the net farm returns at the previous levels, the groundwater use would still be lower, implying positive impact of pro rata pricing on sustainability of groundwater use.

The empirical evidence further reinforces the inference drawn by Kumar (2005) that the arguments against pricing are flawed. One dominant argument against price change is that the higher marginal cost of supplying electricity under a metered system, owing to the high transaction cost of metering, could reduce the net social welfare as a result of reduction in: 1) demand for electricity and groundwater in irrigated agriculture, and 2) net surpluses individual farmers could generate from farming. The second argument is that for power tariff to be in the responsive region of power demand curve, prices have to be so high that it may become socially unviable. We will illustrate this point.

It is understood that the aggregate demand for electricity and groundwater in irrigation is a function of the demand rates (electricity and water requirements per unit of land), and the total area under irrigation (Kumar, 2005). The empirical analyses show that the demand for water and energy per unit of land was lower for water buyers due to increase in unit price of water and energy. However, the net income surpluses from every unit of water and energy increased. Also, the net aggregate return was higher in Bihar and UP. Although the net returns per unit of land were marginally lower for farmers who paid on pro rata basis in north Gujarat, this is not a concern as in water-scarce regions farmers would not face land constraints in maximizing farm returns. With higher water productivity (Rs/m^3), they would be able to maintain the same level of net farm return as in the past, but with much smaller amounts of water.

The increase in income surplus from every unit of water and energy will be more under pro rata pricing because there is no need for regulating power supply, which is done by state electricity boards for restricting revenue losses through subsidy at the macro level, under pro rata pricing, whereas it is compulsory under flat rate system of pricing. Now, if one considers the positive externalities on the society due to energy and water saving resulting from their efficient use, the net social welfare would be even more. Groundwater-scarce regions such as western and northwestern India and south Indian peninsula (Kumar et al., 2010) are also facing severe energy shortages, and hence such welfare effects are likely to be high.

Introducing electricity metering in the farm sector: Technological changes

The importance of metering electricity and charging on pro rata basis for both cost recovery and improving energy efficiency is well recognized by the SEBs and policymakers. Metering and pro rata pricing reduces the unaccounted for losses in electricity distribution, improves the financial working of the SEBs and reduces the overall power deficits. But for almost 2 decades, the SEBs were toiling with the idea of carrying out metering of farm-power connection in a way that makes it foolproof as well as cost effective. The problem was the rampant tampering of meters in rural areas, and malfunctioning meters. Today, technologies exist not only for metering but also for controlling energy consumption by farmers (Kumar and Amarasinghe, 2009; Zekri, 2008). The prepaid electronic meters, which are operated through scratch cards and can work on satellite and Internet technology, are ideal for remote areas to monitor energy use and control groundwater use online from a centralized station (Zekri, 2008). It is important to note here that over the past 7–8 years, there has been a remarkable improvement in the quality of services provided by Internet and mobile (satellite) phone services, especially in the rural areas, with a phenomenal increase in the number of consumers. As Zekri (2008) notes, such technologies are particularly important when there are large numbers of agro wells, and the transaction cost of visiting wells and taking meter reading is likely to be very high. It is inevitable that they will be adopted in rural India.

Pilferage of electricity through manipulation of pump capacity can be prevented through the use of prepaid meters. They can be operated through tokens, scratch cards, magnetic cards or recharged digitally through Internet and SMS. It helps electricity companies restrict the use of electricity. A company can decide on the 'energy quota' for each farmer on the basis of reported connected load and total hours of power supply, or sustainable abstraction levels per unit of irrigated land. But for implementing this, databases of the connected load and coordinates of the farmers, and field data to assess sustainable withdrawal levels, among others, are required. Farmers can pay and obtain activation codes through mobile SMS (Zekri, 2008).

Restricting energy use of farmers for pumping groundwater is analogous to rationing water allocation for irrigation volumetrically. As Kumar (2005) has shown, when water allocation is rationed in volumetric terms, farmers allocate the available water to economically more efficient uses. Hence, restricting energy use will have a positive impact on efficiency of groundwater use by all categories of farmers. But in such cases, it is important that the consumers are informed about their energy quota (in kWh), and the approximate number of hours for which they could pump water from their wells using this quota, well in advance of the agricultural season. Such information would help them choose the crops depending on the availability of power over the entire crop year. Here again, the energy quota will have to be decided on the basis of the geo-hydrological environment prevailing in the area, which determines the energy use per unit volume of groundwater pumping, and the optimum irrigation requirements.

Power supply policies will also impact on land and water productivity in irrigated farming. Unlimited power supply with stable voltage will ensure better quality of irrigation water, than will restricted power supply with voltage fluctuations. As some studies had indicated, the returns from irrigation are highly elastic to its quality (Kumar and Singh, 2001; World Bank, 2001).

Hence, various policy options are possible with regard to energy supply, metering and power pricing, enabled by the use of prepaid meters. Five of them are presented in Figure 6.5. Under all the options except Options 3 and 5, farmers would pay higher tariff for electricity. But they would also gain in terms of improved quality of power. As a study by World Bank (2001) in Haryana showed, many farmers understand that ability of state electricity boards to offer improved electricity service depends on higher tariffs and metering. Fewer understand the need to invest in more efficient pump sets. To ensure continued and increasing support from the farming lobby for the reforms that are put into action, policymakers must clearly communicate to them that the new strategy balances higher costs of power with improved service performance over a timeframe. The likely outcomes and impacts of these policy interventions are also summarized in Figure 6.5.

As Figure 6.5 suggests, Option 3 is easily implementable to manage the groundwater-energy nexus in agriculture. Here, the energy quota is decided on the basis of the connected load of the farmer, which means that those who have large

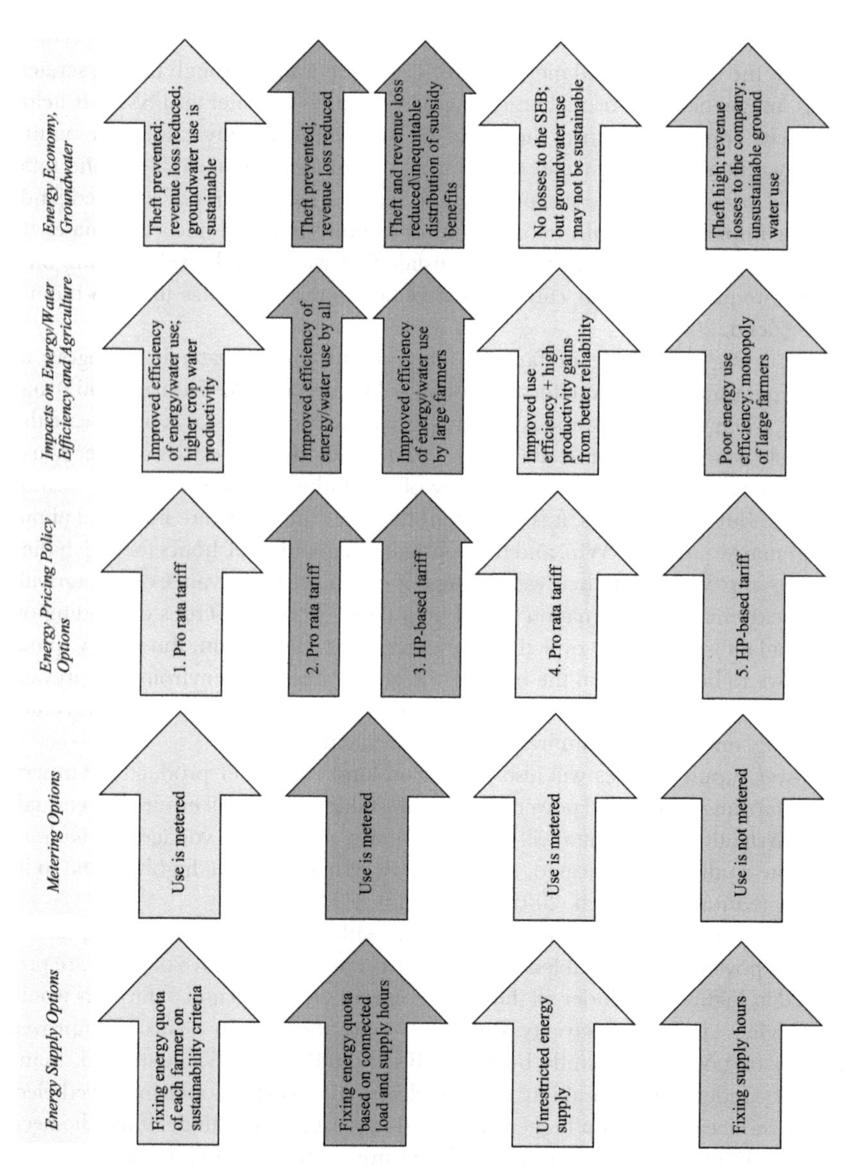

Figure 6.5 Different Modes of Pricing and Expected Outcomes Under Different Energy Supply Regimes

capacity pump sets would be entitled to large quantities of electricity. Therefore, equity in access to groundwater and water use efficiency in small farms could still be an issue. One way to overcome this issue is to fix higher charges per HP of connected for farmers who enjoy large quotas. Option 2 is slightly difficult, but would conserve groundwater also. Option 1 is best for co-management of groundwater and electricity and would address the issues of equity, efficiency and sustainability. But implementing this requires great political will, as rights of farmers to use groundwater would be regulated by this policy intervention. Governments can offer subsidies for meters if farmers are willing to go for Options 1 and 2 on account of the positive welfare effects. The major effects are reduced depletion of groundwater affected by efficient use of the pumped water, and reduced carbon footprint in agriculture due to reduction in use of electricity for groundwater pumping affected by reduced draft and improved pump efficiency. For implementing any of these options, it is necessary that SEBs set up computerized database of all agro wells, comprising their latitude and longitude, physical characteristics and land use data.

Conclusion and policy implications

On the one hand, both electricity consumption and energy subsidy in agriculture are steadily going up in India. On the other, the groundwater draft for irrigation is also increasing, causing widespread problems. Pro rata pricing of electricity coupled with rationing of the energy supply based on groundwater resource sustainability criteria is the best option for co-management of electricity and groundwater, and for reducing the revenue losses to the State Exchequer. This would address the issue of equity, efficiency and sustainability of groundwater use, while improving the energy economy. The second best option would be pro rata pricing with control of electricity supply on the basis of connected load and supply hours, which will positively impact on efficiency of energy/groundwater use and revenue gains for SEBs, although it will not address the issue of sustainable groundwater use. Metering of energy use using prepaid meters is a prerequisite for implementing these options.

Even under the present scenario with several technological advancements changing the communication networks in rural areas, there are some initial transaction costs in introducing prepaid meters. But they will be much less as compared to the past when SEBs had to take meter readings manually from hundreds of thousands of individual farmers in remote rural areas. Also, there are processes involved in implementing the systems such as fixing the energy quota of individual farmers, generating databases of well owners, and providing extension services to the farmers for effectively using the new technologies such as prepaid meters, apart from informing them about their energy quotas. All these would take time, technical and human resources and finance. But there are huge opportunity costs of not doing this, in the form of low agricultural productivity, and threat to sustainability of groundwater resources and livelihoods of millions of farm households. On the other hand, there are economic benefits of following the new system of supplying electricity, metering and charging higher tariff such as improved productivity of use of electricity and water, greater agricultural outputs, increased revenue for the state electricity boards from the farm sector, improved financial viability of the power sector and reduced carbon footprints in agriculture. All these can be done without adversely affecting the economic viability of irrigated production.

Notes

1. This chapter heavily draws from the paper titled 'Inducing the Shift From Flat-Rate or Free Agricultural Power to Metered Supply: Implications for Groundwater Depletion and Power Sector Viability in India' published in *Journal of Hydrology, 409* (2011), pp. 382–394, authored by M. Dinesh Kumar, Christopher A. Scott and O. P. Singh.
2. As per the authors' estimates based on data provided in GOI (2011), the power consumption in agriculture went up from 102.2 billion units to 136.46 billion units during 2007–08 to 2011–12.
3. This is based on the formula that producing one unit of electricity through fossil fuel burning would emit 0.26 kg of carbon (Nelson and Robertson, 2008).
4. Due to interruptions in power supply accompanied by poor quality of power, farmers do not have absolute control over irrigation water. Under this situation, they show increasing tendency to over-irrigate the crops when electricity is available. Water delivery often does not coincide with the critical stages of crop water demand. The result is that they are getting less output per unit of irrigation water.
5. The term 'well-irrigated farm/s' is used interchangeably with 'well command/s.'

References

Amarasinghe, U. A., Sharma, B. R., Aloysius, N., Scott, C., Smakhtin, V. and de Fraiture, C. (2005). *Spatial Variation of Water Supply and Demand Across River Basins of India*. Research Report 83. Colombo, Sri Lanka: International Water Management Institute.

Bhatia, B. and Gulati, M. (2004). Reforming the Power Sector: Controlling Electricity Thefts and Improving Revenue. *Public Reforms for the Private Sector*. Washington, DC: The World Bank.

Central Statistics Office (2012). *Energy Statistics 2012* (19th Issue). New Delhi: Central Statistics Office, Ministry of Statistics and Programme Implementation, Government of India.

Garduño, H. and Foster, S. (2010). Sustainable Groundwater Irrigation Approaches to Reconciling Demand With Resources. Strategic Overview Series No. 4. The World Bank, Washington, DC.

Giordano, M. and Villholth, K. (2007). The Agricultural Groundwater Revolution: Setting the Stage. In M. Giordano and K. G. Villholth (Eds.), *The Agricultural Groundwater Revolution: Opportunities and Threats to Development* (Comprehensive Assessment of Water Management in Agriculture). Oxfordshire: United Kingdom: CABI Publishing.

Government of India (2002, May). Annual Report on the Working of State Electricity Boards and Electricity Department – 2001–02, Planning Commission (Power and Energy Division), Government of India. Retrieved fromhttp://planningcommission.nic.in/

Government of India (2005). *Dynamic Ground Water Resources of India*. New Delhi: Central Ground Water Board, Ministry of Water Resources, Government of India.

Government of India (2011, October). *Annual Report 2011–12 of the Working of the State Power Utilities and Electricity Departments*. New Delhi: Planning Commission, Government of India.

Grey, D. and Sadoff, C. (2007). Sink or Swim: Water Security for Growth and Development. *Water Policy, 9*(6), 545–571.

Howes, S. and Murugai, R. (2003, April 19). Incidence of Agricultural Power Subsidies: An Estimate. *Economic and Political Weekly*.

Kumar, M.D. (2005). Impact of Electricity Prices and Volumetric Water Allocation on Groundwater Demand Management: Analysis From Western India. *Energy Policy, 33*(1), 39–51.

Kumar, M.D. (2007). *Groundwater Management in India: Physical, Institutional and Policy Alternatives*. New Delhi: Sage Publications.

Kumar, M.D. and Amarasinghe, U. (Eds.). (2009). *Water Productivity Improvements in Indian Agriculture: Potentials, Constraints and Prospects*. National River Linking Project Series 4, Challenge Program on Water and Food. Colombo: International Water Management Institute.

Kumar, M.D., Malla, A.K. and Tripathy, S. (2008b). Economic Value of Water in Agriculture: Comparative Analysis of a Water-Scarce and a Water-Rich Region in India. *Water International, 33*(2), 214–230.

Kumar, M.D., Narayanamoorthy, A. and Sivamohan, M.V.K. (2010). *Pampered Views and Parrot Talks: In the Cause of Well Irrigation in India*. Occasional Paper #1. Hyderabad: Institute for Resource Analysis and Policy.

Kumar, M.D. and Patel, P.J. (1995). Depleting Buffer and Farmers' Response: Study of Villages in Kheralu, Mehsana, Gujarat. In M. Moench (Ed.), *Electricity Prices: A Tool for Ground Water Management in India?* Ahmedabad: VIKSAT-Natural Heritage Institute.

Kumar, M.D., Scott, C.A. and Singh, O.P. (2011) Inducing the Shift from Flat-Rate or Free Agricultural Power to Metered Supply: Implications of Groundwater Depletion and Power Sector Viability in India. *Journal of Hydrology, 409* (2011), 382–394.

Kumar, M.D. and Singh, O.P. (2001). Market Instruments for Demand Management in the Face of Scarcity and Overuse of Water in Gujarat. *Water Policy, 3*(5), 387–403.

Kumar, M.D., Singh, O.P., Samad, M., Purohit, C. and Malkit, S.D. (2008a). *Water Productivity of Irrigated Agriculture in India: Potential Areas for Improvement*. Proceedings of the 7th Annual Partners Meet, IWMI-Tata Water Policy Research Program "Managing Water in the Face of Growing Scarcity, Inequity and Declining Returns: Exploring Fresh Approaches," ICRISAT Campus, Patancheru, April 2–4.

Kumar, M.D., Singh, O.P. and Singh, K. (2001). *Groundwater Development and Its Socioeconomic and Ecological Consequences in Sabarmati River Basin*. Monograph 2. Anand: INREM Foundation.

Mukherji, A., Das, B., Majumdar, N., Nayak, N.C., Sethi, R.R., Sharma, B.R. and Banerjee, P.S. (2009). Metering of Agricultural Power Supply in West Bengal: Who Gains and Who Loses? *Energy Policy, 37*(12), 5530–5539.

Nelson, G.C. and Robertson, R. (2008). *Estimating the Contribution of Groundwater Irrigation Pumping to CO2 Emissions in India*. Washington, DC: International Food Policy Research Institute, Washington.

Reddy, D.C.R.V. (2009). Water Pricing as a Demand Management Option: Potentials, Problems and Prospects. In R.M. Saleth (Ed.), *Promoting Irrigation Demand Management in*

India: Potentials, Problems and Prospects (pp. 25–46). National River Linking Project Series 3. Colombo: International Water Management Institute.

Saleth, R. M. (1997). Power Tariff Policy for Groundwater Regulation: Efficiency, Equity and Sustainability. *Artha Vijnana, XXXIX*(3), 312–322.

Schiffler, M. (1998). The Economics of Groundwater Management in Arid Countries: Theory, International Experience and a Case Study of Jordan. London: GDI Book Series No. 11.

Scott, C. A. (2011). The Water-Energy-Climate Nexus: Resources and Policy Outlook for Aquifers in Mexico. *Water Resources Research, 47*. doi: 10.1029/2011WR010805

Scott, C. A. and Shah, T. (2004). Groundwater Overdraft Reduction Through Agricultural Energy Policy: Insights From India and Mexico. *International Journal of Water Resources Development, 20*, 149–164.

Scott, C.A. and Sharma, B. (2009). Energy Supply and the Expansion of Groundwater Irrigation in the Indus-Ganges Basin. *International Journal of River Basin Management, 7*(2), 119–124.

Shah, T. (1993). *Water Markets and Irrigation Development: Political Economy and Practical Policy.* Bombay: Oxford University Press.

Shah, Z. and Kumar, M. D. (2008). In the Midst of the Large Dam Controversy: Objectives, Criteria for Assessing Large Water Storages in the Developing World. *Water Resources Management, 22*, 1799–1824.

Sharma, K. D. (2009). Groundwater Management for Food Security in India. *Current Science, 96*(11), 1444–1447.

Singh, O. P. (2004). *Water Productivity of Milk Production in North Gujarat, Western India.* Proceedings of the 2nd Asia Pacific Association of Hydrology and Water Resources (APHW) Conference, Vol. 1, pp. 442–449.

US Energy Information Administration. (2011). Country analysis briefs – India. *Independent Statistics and Analysis,* updated November 21, 2011.

Vashishtha, P. S. (2006a). *Input subsidy in Andhra Pradesh agriculture.* Unpublished manuscript.

Vashishtha, P. S. (2006b). *Input subsidy in Punjab agriculture.* Unpublished manuscript.

World Bank (2001). *India: Power Supply to Agriculture.* South Asia Region, Washington: Author.

World Bank (2010). *Deep Wells and Prudence: Towards Pragmatic Action for Addressing Groundwater Over-Exploitation in India.* Washington DC: Author.

Zekri, S. (2008, March). Using Economic Incentives and Regulations to Reduce Seawater Intrusion in the Batinah Coastal Area of Oman. *Agricultural Water Management, 95*(3).

7 Diesel price hikes and farm distress in the fossil fuel-dependent agricultural economy

Myth and reality[1]

M. Dinesh Kumar, O. P. Singh and M. V. K. Sivamohan

Introduction

The unprecedented fall in the poverty levels of Asian countries during the recent decades mirrors the large contribution made by the green revolution to the structural transformation (Ravallion and Chen, 2004). The poverty reduction ability of agricultural growth in India was well demonstrated by Datt and Ravallion (1998). Though this is being questioned in the light of the declining contribution of agriculture to national economies of many south Asian countries (see Byerlee et al., 2005), this argument in development thinking that agricultural growth can reduce poverty is still applicable to many backward regions that are still agrarian.

Paradoxically, water-abundant eastern India has the lowest agricultural growth in the country (Evenson et al., 1999) and the highest rate of rural poverty. The economic scarcity of water is the key water-management issue in agriculture (Kumar, 2003), and a cause for agricultural backwardness (Shah, 2001). Due to inadequate rural electrification, high cost of diesel engines and the small size of land holdings, investment in irrigation is very low. The resource-poor, small and marginal farmers pay exorbitant prices for the water they buy from well owners, making irrigated agriculture an unattractive proposition (Kumar, 2007).

Of late some researchers have argued that the hikes in diesel prices had badly hit agricultural growth in regions with poor rural electrification, including eastern India, causing widespread farm distress in the countryside (Mukherjee, 2007; Shah, 2007). This is based on the premise that regions such as eastern UP, West Bengal, Assam and Bihar depend heavily on diesel power for lifting groundwater, and increase in price of diesel is likely to raise irrigation costs, significantly reducing farm incomes, as agricultural productivities are already very low there. Over the period of 16 years from 1990 to 2006, the price of diesel has also gone up from Rs. 5 per literto Rs. 34.84 per liter.

The limited empirical work on the impact of energy price hikes on irrigated farming is based on respondent surveys (see, for instance, Shah, 2007). Such works have little relevance for policy formulation in the sense that the perceived 'impacts of price changes' are an outcome of the whole range of changes happening with the farming system, including that on the market front. Such analyses fail to segregate the response of farmers to input price changes, and their subsequent implications for prospects of farming, especially of small and marginal farmers. They also fail to nullify the effect of the market, particularly changes in output prices.

A conceptual framework for analyzing the impact of energy price on farming enterprise

The rise in diesel price might affect the diesel well owners and water buyers in diesel well commands in different ways. First of all, the rising diesel price (Rs/liter) would raise the marginal cost of using irrigation water for crop production unless the farmers put in place the systems to improve efficiency of the pump that reduce the diesel consumption per hour of pumping, and increase the well outputs. This would encourage the well owners to improve the efficiency of use of irrigation water in the field to minimize the cost of irrigation. This can result in lower irrigation dosage, higher yield, higher physical productivity of water (kg/m^3) and perhaps higher water productivity in economic terms also (Rs/m^3), with net returns remaining the same.

But there are crops for which the cost of irrigation itself constitutes the lion's share in the production cost. In such instances, with the diesel price hike, irrigation could become unaffordable even with efficiency improvement, when compared with the gross returns from the crop. This would eventually make the crop economically not viable. In such situations, the farmers might respond by replacing the existing crops with new crops that yield much higher income returns (Rs/ha of land) so as to offset the increase in cost of irrigation water. This can also include crops that are highly water intensive. Hence, analysis should be based on farming system returns rather than return from crops.

The impact of rising diesel price on water buyers is more complex. The reason is that over time, the monopoly price of water can also tilt in favor of water buyers due to an increase in the number of privately owned wells. In Muzaffarpur in Bihar, for example, the monopoly rent for groundwater pumping fell by half over 30 years (Kishore, 2004; Kumar, 2007). In such eventuality, the price paid by water buyers may not rise in proportion to the increase in cost of production of water. Therefore, taking the actual price paid by water buyers at different time periods would give a distorted picture of the impact of diesel price on farmers.

The market price of many agricultural products also can change significantly over time. For instance, the price of many cereals did not increase in real terms till recently, and on the contrary had reduced in certain cases. This might have affected the returns from farming adversely. Hence, comparing the time series data on income returns from farming would distort the picture with respect to the impact of input prices by wrongly attributing the declining returns to the increase in input costs. In order to capture the 'response to and impact of input price changes,' it is essential to compare the farming systems under different price regimes for irrigation water. The major attributes to be covered in such analysis include: 1) irrigation dosage, 2) physical productivity of water in crops, 3) net returns from crops and livestock, 4) net water productivity in economic terms for crops and livestock, and 5) the returns and water productivity in farming.

Objectives, hypothesis and methodology

The overall objective of the study is to analyze how the small and marginal farmers in water-abundant regions respond to a diesel price hike, and to assess the overall impact on the economic prospects of farming. More specifically, it aims to analyze the actual change in cost of irrigation water for different factors due to a rise in diesel prices; analyze the response of diesel well owners to a hike in irrigation costs and its overall impact on economic prospects of farming; and study the response of water buyers in diesel well commands to price hike, and its overall impact on economic prospects of their farming.

It is hypothesized that increase in cost of diesel would encourage farmers to use water and other inputs for crop and livestock production more efficiently; but they might also adopt high-valued crops that require more water, to sustain the net returns from crop production.

The study area covers locations that include the eastern plain region of Uttar Pradesh and south Bihar plains. The average annual precipitation in the eastern plain region of Uttar Pradesh is about 1,025 mm. The region's climate is dry sub-humid to moist sub-humid. The soil types in this sub-zone are light alluvial and calcareous clay. Patna district lies within the south Bihar plains region of Bihar state and receives average normal annual rainfall of 1,103 mm, and the climate condition of region is dry to moist sub-humid. The soil types found in the region are old alluvium sandy loam to clayey and the larger areas are *Diara* (flood plains) and *Tal* (low-lying areas other than flood plains).

The details of sample selected for the study are given in Table 7.1.

At the outset, an analysis of the actual increase in cost of irrigation water for the diesel well owners and water buyers over time is made on the basis of the time series data on cost of diesel, energy (diesel) use per hour of pumping and the well outputs. This is adjusted for inflation. The price water buyers pay in hourly terms

Table 7.1 Sampling Procedure and Sample Size

Name of the Regions	*Name of the District*	*Water Buyers in Electric Well Command*	*Diesel Pump*		*Total Sample Size*
			Well Owner	*Water Buyer*	
Eastern UP	Varanasi and Mirzapur	60	60	60	180
South Bihar	Patna	60	60	60	180
Total		**120**	**120**	**120**	**360**

is converted into unit charges using data on well outputs, and then adjusted for inflation to obtain the prices in real terms.

The potential response of diesel well owners to increase in irrigation cost is analyzed by comparing the cropping pattern; irrigation water dosage; and productivity of irrigation water in crops, dairying and at the farm level of the diesel well owners with that of water buyers from electric well owners. This is in view of the fact that the water buyers in electric well commands pay much less for irrigation water as compared to diesel well owners, and water buyers in diesel well command. Subsequently, the potential response of water buyers to a rise in irrigation cost is analyzed by comparing the above attributes with water buyers in electric well commands.

The overall economic impact of a rise in irrigation cost on farming enterprise is analyzed by estimating the net return from the entire farm for diesel well owners and their water-purchasing counterparts, and then comparing it with that for water buyers from electric well owners in the same region. Further, based on the actual rise in cost of irrigation water felt by diesel well owners over the years due to diesel price hikes, and the potential response of the irrigators to rising irrigation cost, the actual impact of a diesel price hike on farming enterprises of diesel well owners and their water buyers is assessed.

For the analysis, the net economic return from farming is assessed by adding up the net return from crops and dairy production.

The net return from crop production NI_{crop} for those crops having byproducts that do not get used for dairying is estimated as:

$$NI_{Crop} = \left[(Y_{MP} * FHP_{MP}) + (Y_{By\text{-}P} * FHP_{By\text{-}P})\right] - C_{Input} \tag{1}$$

Here, Y_{MP} is the yield of main product (kg); FHP_{MP} is the farm harvest price of main product (Rs/kg); $Y_{By\text{-}P}$ is the yield of byproduct (kg) and $FHP_{By\text{-}P}$ is the farm harvest price of byproducts (Rs/kg); C_{Input} is the cost of all inputs used for crop production.

For the estimation of net return NI_{Cropi} for those crops, which have byproducts that are used as an input of milk production (dry fodder), in such situations, we will allocate total cost of production between the main product and byproduct using their ratio of market value. The net return from such crops will be estimated as:

$$NI_{Crop\,i} = \left[(Y_{MP} * FHP_{MP}) - C_{Main\ product}\right] \tag{2}$$

Here, Y_{MP} is the yield of main product (kg); FHP_{MP} is the farm harvest price of main product (Rs/kg); and $C_{Main\ product}$ is the cost of inputs for main production.

The net income NI_{Dairy} from livestock production based on life cycle is estimated as:

$$\begin{aligned} NI_{Dairy} = {} & \left[(Y_{milk} * P_{milk}) + (Y_{Dung} * P_{Dung})\right] \\ & - \left[(P_{gf} * Q_{gf}) + (P_{df} * Q_{df}) + (P_{cf} * Q_{cf}) + P_{oi}\right] \end{aligned} \tag{3}$$

Here, Y_{milk} is the milk yield per animal per annum; Y_{Dung} is the yield of dung per animal per annum; P_{milk} and P_{Dung} are the price of milk and price of dung received by the farmer; Q is the quantum of inputs used per cattle unit per annum. The suffixes *gf*, *df*, *cf* and *oi* stand for green fodder, dry fodder, cattle feed, and other expenses respectively. The price of *gf* and *df* will be the unit cost of production (total input costs divided by total production). The price of *cf* and *oi* are the actual market price.

The net income NI_{farm} at farm level (Rs) would be estimated as:

$$NI_{farm} = \sum_{i=1}^{m} NI_{crop\,i} + \sum_{j=1}^{n} NI_{Dairy} \tag{4}$$

Where, $\sum_{i=1}^{n} NI_{Crop\,i}$ is the net income from all the crops grown by the farmers on his farm and $\sum_{n=1}^{n} NI_{Dairy}$ is the net income from dairy farming.

Estimation of water productivity in dairying and farming systems

The physical water productivity for a given crop (kg/m^3) will be estimated using data on crop yield and the estimated volume of water applied for all sample farmers growing that crop. The combined physical and economic water productivity in Rs/m^3 is estimated using data on net returns from crop production in Rs/ha and estimated volume of water in cubic meters. To estimate the net income from a particular crop, the data on inputs for each crop were obtained by primary survey of farmers. This included cost of seed, labor, fertilizer, pesticides and insecticides, irrigation, plowing, harvesting and threshing.

The physical productivity of water in milk production for livestock WP_{Milk} (liters/m^3) can be defined as:

$$WP_{Milk} = \frac{Q_{MP}}{\Delta_{Milk}} \tag{5}$$

Where, Q_{MP} is the average daily milk output by one unit of livestock category over the entire lifecycle (liters/animal/day). Δ_{Milk} is the total volume of water used per animal per day, including the water embedded in feed and fodder inputs, used in dairying for an animal in a day, worked out for the entire animal lifecycle (m^3/animal/day). It is estimated as:

$$\Delta_{milk} = \frac{Q_{cf}}{WP_{cf}} + \frac{Q_{df}}{WP_{df}} + \frac{Q_{gf}}{WP_{gf}} + \Delta_{DW} \qquad \text{(6) (Kumar, 2007)}$$

Where Q_{cf}, Q_{df} and Q_{gf} are the average quantities of cattle feed, dry fodder and green fodder used for feeding a livestock unit per day (kg/animal/day); WP_{cf}, WP_{df} and WP_{gf} are the physical productivities (kg/m^3) of cattle feed, dry fodder and green fodder, respectively; Δ_{DW} is the daily drinking water consumption by livestock (m^3/day). It is the average volume of water required by a dairy animal per day over its entire lifecycle, including the water embedded in feed and fodder.

Q_{cf}, Q_{gf}, Q_{df} and Δ_{DW} for a given category of livestock would be estimated for the entire lifecycle of the animal from the following: 1) weighted average of the average daily figures of these inputs for each season for animals in different stages of the lifecycle, viz., calving, lactation stage, dry stage; and 2) the time period in each stage of animal lifecycle for that category of livestock.

Since all the farmers in the sample may not have animals that represent all the different stages of the lifecycle in a particular category of livestock at a given point of time, the average values of inputs are worked out as value of above-mentioned variables for the sample farmers. Likewise, the average values of physical productivity of water in green fodder and dry fodder are used for estimation. Q_{MP} (liter/animal/day) is estimated from: 1) the weighted average of average daily figures of milk yield for different seasons and 2) the ratio of time period in lactation and the average lifespan of the animal in that category.

WP_{gf} and WP_{df} are estimated by taking their respective quantities and the volume of water required for growing that crop. In the case of byproducts of crops used as fodder, the water used for growing that crop is allocated as the main product and byproduct in proportion to the market prices of the respective (Singh, 2004).

The net return of milk production, NR_{milk} (Rs/animal/day) is estimated using values of Q_{MP}, the price of milk (Rs/liter) and the cost of production of the average amount of cattle inputs required in a day (Rs/animal/day) estimated for the entire animal lifecycle as proposed by Singh (2004) and Kumar (2007). It is important to mention here that with import of green or dry fodder in a farm, the cost of fodder input could also go up. This in turn would affect net water productivity in dairying WP_{dairy} (Rs/m^3). It can be estimated as:

$$WP_{dairy} = \frac{NR_{milk}}{\Delta_{milk}} \quad (7) \text{ (Kumar, 2007)}$$

In the case of purchase of inputs, market price is used. If the inputs are from the farmers' own fields, the actual cost of production is estimated. If farmer uses crop byproducts for dairying, the total cost of production of the given crop is allocated among the main product and byproduct on the basis of the potential revenues that can be earned from their sale. The quantity of inputs (feed and dry and green fodder) and milk outputs are worked out for the entire animal lifecycle and not on the basis of the actual use of inputs and milk yield at the point under consideration.

The total volume of water used for milk production annually by one unit of livestock V_{dairy} (m^3/animal/annum) is estimated by dividing the total annual milk production by one unit of livestock (Q_{AMP}) by the physical productivity of water in milk production (WP_{milk}).

The water productivity of the farm WP_{farm} (Rs/m^3), including crops and dairy, is estimated as:

$$WP_{farm} = \frac{\sum_{i=1}^{m} WP_{crop\,i} V_{crop,i} + \sum_{j=1}^{n} WP_{dairy} V_{dairy,j} N_j}{\sum_{i=1}^{m} V_{crop,i} + \sum_{j=1}^{n} V_{dairy,j}} \quad (8)$$

Here, $WP_{crop,i}$ is the water productivity of main product of crop *i*; $V_{crop,i}$ is the total volume of water used for crop *i*; $WP_{dairy,j}$ is water productivity in dairy production for livestock type *j*; and $V_{dairy,j}$ is the volume of water used for dairy production per animal for livestock category *j*. N_j is the total number of livestock in category *j*.

Results and discussions

Impact of diesel price hikes on cost of pumping well water and water prices in the market

The actual cost of groundwater pumping and the price at which water is traded in the diesel well commands of eastern UP and south Bihar are given in Table 7.2. It shows that in the case of south Bihar villages, the average cost of pumping went up by nearly 300% from Rs. 0.41/m^3 to 1.16/m^3, whereas the selling price of water went up by only 90% from Rs. 1.16/m^3 to Rs. 2.11/m^3 from 1990 to 2006. This means that the monopoly price ratio had declined from 2.8 to 1.3. In the case of eastern UP villages, the average cost of pumping went up by nearly 280% from Rs. 0.47/m^3 to Rs.1.64/m^3, whereas the sale price of water went up by 170%. Here, the decline in monopoly price of water is not very sharp, as the rate of growth in the number of diesel wells and pump sets has been less in eastern UP.

The above estimates are useful for analyzing the changes in monopoly hold of well owners, but have limited use when it comes to analyzing the changes in price of irrigation water in real terms over time. This is because of the time value of money. If one wants to see how the price has changed in real terms, it has to be adjusted to the inflation rates. Using an annual inflation rate of 7% for the period from 1990 to 2006, analysis shows that in the case of eastern UP,

Table 7.2 The Cost of Pumping and Sale Price of Groundwater in Diesel Well Commands of Eastern UP and South Bihar Villages

Name of the Region	*Price Details*	*Cost of Pumping Groundwater (Rs/m³)*				*Selling Price of Groundwater (Rs/m³)*			
		1990	*1995*	*2000*	*2006*	*1990*	*1995*	*2000*	*2006*
South Bihar Plains	Average	0.41	0.51	0.95	1.60	1.16	1.40	1.75	2.11
	Minimum	0.22	0.31	0.75	1.38	0.90	1.10	1.44	1.81
	Maximum	0.95	1.04	1.46	2.09	1.67	1.88	2.08	2.50
Eastern UP	Average	0.47	0.56	1.00	1.64	0.98	1.31	1.63	2.67
	Minimum	0.14	0.21	0.54	1.02	0.84	1.11	1.39	1.96
	Maximum	1.51	1.62	2.10	2.82	1.62	2.34	2.52	3.51

Note: Here we assumed that the diesel consumption, discharge rate of the pump and hours of diesel pump running per year is same for 2006.

Source: Authors' own analysis based on primary data

the cost of irrigation in real terms went up by 18%, whereas the actual price of irrigation water from diesel well owners dipped by 7.5%. In south Bihar, while the cost of irrigation in real terms went up by 32%, the price of irrigation water dipped by 38%.

Differential cost of groundwater for irrigation across different categories of farmers

The average cost of pumping irrigation water for the diesel pump owners in Eastern UP is Rs. 1.38/m^3 and the range is Rs. 0.99/m^3 to Rs. 2.04/m^3. The average price at which diesel well owners sell water is Rs. 2.81/m^3 and the individual values range from Rs. 2.07/m^3 to Rs. 3.63/m^3. Against this, the average price at which irrigation water is sold by electric pump owners is Rs. 0.65/m^3 and the individual values range from Rs. 0.52/m^3 to Rs. 0.84/m^3.

The average price at which groundwater is being sold by electric well owners in South Bihar is Rs. 0.70/m^3 and the individual values range from Rs. 0.31/m^3 to Rs. 0.92/m^3. The average cost of pumping groundwater using diesel pumps is Rs.1.87/m^3 and individual values range from Rs. 1.41/m^3 to 2.93/m^3. The average price at which groundwater is being sold by diesel pump owners is Rs. 2.15/m^3 and the individual values range from Rs. 1.84/m^3 to Rs. 2.42/m^3.

Irrigation, net return from crops and crop water productivity

The estimates of irrigation dosage, water productivity in physical and economic terms for selected crops for diesel well owners and their water buyers in eastern UP are presented in Table 7.3. Comparison of the estimates for diesel well owners and their water-buying counterparts shows the following: 1) the average depth of irrigation is slightly lower for water buyers in diesel well command as compared to their well-owning counterparts and 2) the water buyers in diesel well commands secure higher physical productivity of water and water productivity in economic terms for all the crops.

Comparison between diesel well owners and water buyers in electric well commands, however, shows a different trend. The average dosage of irrigation is much lower for farmers who buy water from electric well commands as compared to diesel well owners, in spite of the fact that they are confronted with much lower marginal cost of irrigation. Further, the water buyers in electric well command secure much higher values of water productivity in both physical and economic terms as compared to diesel well owners. This is due to the fact that the electric well commands are located in the flood plains of the river, with high soil moisture content and fertile soils. These reduce not only the irrigation water requirement of the crops, but also the need for fertilizer inputs, minimizing the input costs. As a result, the irrigation dosage and water productivity values are higher for water buyers in electric well commands.

As regards the net income from crop production, the comparative figures for eastern UP are presented in Figure 7.1. Here, the electric well water buyers secure higher income per ha in paddy, wheat, pea and gram as compared to farmers in diesel well commands. Whereas, the diesel pump water buyers secure higher net income in bajra (pearl millet), mustard and linseed.

Comparison of the estimates of irrigation dosage, water productivity (physical and economic) for selected crops for diesel well owners and their water buyers in south Bihar shows a similar trend as that of eastern UP. But comparison of corresponding figures between diesel well owners and water buyers in electric well

Table 7.3 Water Use, Physical Productivity of Water and Net Water Productivity in Economic Terms, Eastern Uttar Pradesh

Name of the Crops	*Diesel Pump – Owner*			*Diesel Pump – Water Buyer*			*Electric Pump – Water buyer*		
	Depth of Irrigation (cm)	*WP (kg/ m³)*	*Net WP (Rs/ m³)*	*Depth of Irrigation (cm)*	*WP (kg/ m³)*	*Net WP (Rs/ m³)*	*Depth of Irrigation (cm)*	*WP (kg/m³)*	*Net WP (Rs/m³)*
Paddy	15.53	1.86	2.62	9.09	2.39	2.92	3.61	2.29	3.64
Sesame	2.29	0.89	17.39	1.14	0.88	17.72	0.57	1.25	9.58
Sugarcane	–	–	–	–		–	0.57	10.62	8.11
Bajra	2.29	3.43	7.47	1.33	4.41	17.83	1.43	4.05	10.52
Wheat	12.74	2.57	6.22	8.33	3.50	7.80	2.93	2.63	7.57
Potato	3.70	7.23	17.87	2.29	7.40	–	2.91	5.96	9.58
Pea	3.40	1.56	12.19	1.67	1.74	12.36	1.33	2.14	14.95
Gram	4.16	1.58	15.33	1.99	1.82	17.78	0.36	1.62	31.12
Mustard	2.70	1.56	10.87	1.44	1.15	11.99	1.20	1.39	11.44
Linseed	3.43	1.36	13.70	1.03	1.53	16.77	–	–	–

Source: Authors' own estimate based on primary data

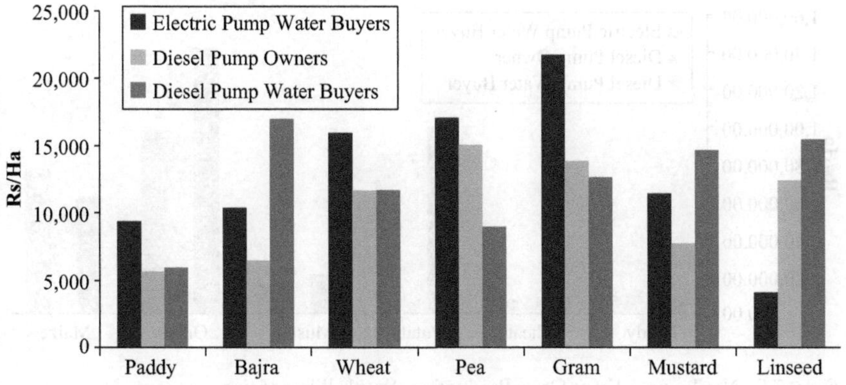

Figure 7.1 Net Income From Crop Production, Eastern UP

command shows the same trend as that in the case of diesel well owners and their water-buyer counterparts in the same region. The irrigation dosage and water productivity (in both physical and economic terms) are higher for diesel well owners who have to pay a much higher price for irrigation water in volumetric terms, as compared to those who buy water from electric well owners, unlike what was found in the case of eastern UP.

The comparative figures of net income from crop production for south Bihar are presented in Figure 7.2. In the south Bihar plains, the electric well water buyers obtain higher net income per ha in wheat, potato, mustard and onion crops, whereas diesel pump water buyers get higher net income per ha in maize cultivation. The diesel pump owners are receiving their second highest per hectare net income from onion, mustard, potato and wheat and highest per hectare net income from the paddy crop.

Net income and water productivity in milk production

The net income from milk production is dependent on the amount of fodder and feed provided to the dairy animals, cost of production of these inputs and milk yield and the price of these outputs. The determinants of water productivity in milk production are the milk yield and its price, the total amount of water embedded in the animal feed and fodder and the cost of production of animal feed and fodder.

Feed and fodder use

Table 7.4 presents the estimates of average quantum of green fodder, dry fodder and animal feed provided to the three different types of livestock, viz., buffalo, crossbred cow and indigenous cow, for diesel well owners; the farmers who

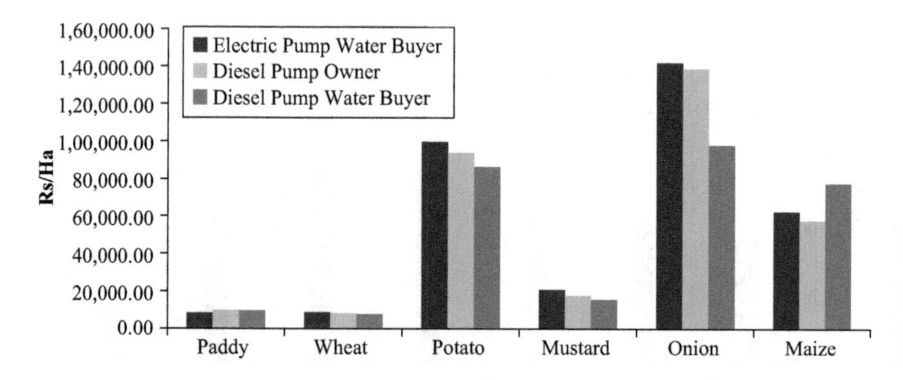

Figure 7.2 Net Income From Crop Production, South Bihar Plains

Table 7.4 Average Feed and Fodder Used Based on Lifecycle of Animal, Eastern UP

	Feed and Fodder Use (kg/day/animal)								
	Diesel Pump – Owner			*Diesel Pump – Water Buyer*			*Electric Pump – Water Buyer*		
	Buffalo	*CB Cow*	*Ind. Cow*	*Buffalo*	*CB Cow*	*Ind. Cow*	*Buffalo*	*CB Cow*	*Ind. Cow*
Green Fodder	19.6	24.3	17.1	19.2	16.7	13.4	13.8	19.5	14.8
Dry Fodder	15.1	17.5	13.2	17.1	19.6	19.1	8.9	12.7	9.3
Concentrate	1.5	1.5	1.4	1.3	1.8	1.3	1.0	1.8	1.2

Source: Authors' own analysis based on primary data

purchase irrigation water from them; and those who purchase water from electric pump owners in eastern UP villages. Comparison across the different categories of farmers shows that generally, the dairy inputs are highest for water buyers of diesel well owners, followed by diesel well owners, and lowest for water buyers in electric well commands. The exception is for crossbred cows, in which case the green fodder input is much lower for water buyers of diesel well command as compared to other categories of farmers.

Comparisons of corresponding estimates for south Bihar villages show no general trend, unlike what has been found in the case of eastern UP. Nevertheless, for buffalo, the quantum of green and dry fodder input was highest for diesel well owners there.

Water use for milk production, physical productivity of water and gross water productivity in economic terms

The estimates of total water use, physical productivity of water in milk production (liters/m^3) and the gross water productivity in economic terms (Rs/m^3) estimated for eastern UP are presented in Table 7.5. Comparison of the figures of physical productivity of water in milk production shows that the figures are highest for water buyers in electric well commands. Between diesel well owners and water buyers in their command, no major differences were noticeable. As regards gross water productivity in economic terms, the water buyers in electric well commands obtain the highest values in buffalo milk production. Again, no major differences were noticeable between diesel well owners and those who buy water from them in gross water productivity. But, in the case of crossbred cows, diesel well owners secure the highest water productivity, followed by water buyers in diesel well commands, with the lowest productivity secured by water buyers in electric well commands. In the case of indigenous cows, the water buyers in diesel well commands obtain highest water productivity in rupee terms.

Table 7.5 Water Use and Physical Productivity of Water in Milk Production, Eastern UP (m^3/day)

	Diesel Pump – Owner			*Diesel Pump – Water Buyer*			*Electric Pump – Water Buyer*		
	Buffalo	*CB Cow*	*Ind. Cow*	*Buffalo*	*CB Cow*	*Ind. Cow*	*Buffalo*	*CB Cow*	*Ind. Cow*
Total Water Use (m^3)	3.02	3.48	2.68	3.00	3.21	2.64	2.19	3.35	2.38
Milk Production (liters)	2.08	4.01	1.95	2.23	3.23	2.01	2.64	4.08	1.89
WP (Lt/m^3)	0.69	1.15	0.73	0.75	1.00	0.76	1.20	1.22	0.79
Gross WP (Rs/m^3)	11.03	16.13	10.95	11.93	14.06	11.38	12.97	12.31	7.35

Source: Authors' own analysis based on primary data

Table 7.6 Water Use and Physical Productivity of Water in Milk Production, South Bihar (m^3/day)

	Diesel Pump – Owner			*Diesel Pump – Water Buyer*			*Electric Pump – Water Buyer*		
	Buffalo	*CB Cow*	*Ind. Cow*	*Buffalo*	*CB Cow*	*Ind. Cow*	*Buffalo*	*CB Cow*	*Ind. Cow*
Total Water Use (m^3)	4.88	3.93	2.73	3.62	3.18	3.04	4.09	5.36	3.37
Milk Production (liters)	1.69	3.53	1.37	1.68	2.30	1.18	1.86	2.97	0.88
WP (Lt/m^3)	0.35	0.90	0.50	0.46	0.72	0.39	0.45	0.55	0.26
Gross WP (Rs/m^3)	4.85	10.60	7.00	6.50	8.52	5.45	6.35	7.69	3.64

Source: Authors' own analysis based on primary data

The estimates for south Bihar are presented in Table 7.6. Comparison of the figures of physical productivity of water in milk production shows that the figures are highest in the case of water buyers in diesel well commands for buffalo milk; and highest in case of diesel well owners for crossbred cows, and indigenous cows. The water buyers in diesel well commands secure highest gross water productivity in economic terms in buffalo milk production, while diesel well owners secure highest gross water productivity in crossbred cow and indigenous cow milk production. The farmers in electric well commands obtain the lowest figures of physical productivity of water and gross water productivity in economic terms in crossbred cow and indigenous cow milk production.

Expenditure, net income and net water productivity in milk production

Table 7.7 presents the total expenditure in dairy production and net income from dairying per animal per day, and net water productivity in economic terms for different types of livestock. Comparing the estimates for the three different categories of farmers shows some definite trends. For instance, the expenditure on dairy production is highest for water buyers in diesel well commands, who are confronted with highest marginal cost of irrigation water, followed by diesel well owners and finally water buyers in electric well commands. Water buyers in electric well commands secure highest net water productivity in buffalo milk production; diesel well owners obtain highest water productivity in crossbred cow milk production; and water buyers in diesel well commands obtain highest water productivity in production of indigenous cow milk.

But corresponding estimates for south Bihar villages throw up no definite trend vis-à-vis expenditure, incomes and water productivity.

Table 7.7 Net Water Productivity in Economic Terms in Milk Production, Eastern Uttar Pradesh

	Diesel Pump – Owner			*Diesel Pump – Water Buyer*			*Electric Pump – Water Buyer*		
	Buffalo	*CB Cow*	*Ind. Cow*	*Buffalo*	*CB Cow*	*Ind. Cow*	*Buffalo*	*CB Cow*	*Ind. Cow*
Total Expenditure (Rs/day)	17.71	20.12	17.71	20.02	22.73	18.43	12.73	19.80	14.03
Milk Production (Lt)	2.08	4.01	1.95	2.23	3.23	2.01	2.64	4.50	1.89
Income (Rs)	33.34	56.18	29.32	35.74	45.17	30.09	28.44	45.48	17.51
Income From Dung (Rs/day)	0.50	0.50	0.50	0.50	0.50	0.50	0.50	0.50	0.50
Gross Income (Rs/day)	33.84	56.68	29.82	36.24	45.67	30.59	28.94	45.98	18.01
Net Income (Rs/day)	16.13	36.56	12.10	16.22	22.94	12.15	16.21	26.18	3.98
Net WP (Rs/m^3)	5.33	10.50	4.52	5.41	7.14	4.60	7.39	7.82	1.67

Source: Authors' own analysis based on primary data

Impact of differential cost of irrigation on water productivity at the farm level

The figures for water productivity for the crop combinations; water productivity in dairying; and farm-level water productivity for the locations, viz., eastern UP and south Bihar, are provided in Table 7.8. It shows that water productivity figures are higher for the farmers who are confronted with the higher cost of irrigation water, i.e., the water buyers in diesel well commands. The exception is the water buyers in electric well commands. This is due to the inherent advantage with the location.

The higher values of farm-level water productivity for water buyers in diesel well commands is mainly due to the reduced application of irrigation water for the crops, which enhances both physical productivity and water productivity in economic terms for the crops. But the improvement in physical productivity of water does not get converted into higher water productivity in rupee terms in dairy production. This is because the cost of production of dairy inputs is higher for water buyers in diesel well commands and diesel well owners.

Impact of differential cost of irrigation on farm incomes

The impact of cost of irrigation on net returns from farming is insignificant. As Table 7.9 shows, in eastern UP, the water buyers in diesel well commands earn the highest income per unit of land, and the electric well water buyers obtain the lowest net return from unit of land. Table 7.10 shows that in south Bihar, the water buyers in electric well commands earn the highest income per unit of land, followed by water buyers of diesel well commands. The difference in net returns between diesel well owners and their water-purchasing counterparts is also not significant.

Table 7.8 Comparison of Water Productivity at the Farm Level at Different Irrigation Costs in Eastern UP and South Bihar

Name of Region	*Farmer Category*	*Overall Water Productivity in Crops (Rs/m³)*	*Water Productivity in Dairying (Rs/m³)*	*Farm-Level Water Productivity (Rs/m³)*
Eastern UP	WB in electric well	16.72	2.84	9.78
	Diesel well owner	10.35	3.03	6.73
	WB in diesel well	20.06	0.17	10.20
South Bihar	WB in electric well	17.97	1.64	9.81
	Diesel well owner	21.62	0.90	11.27
	WB in diesel well	22.59	0.30	11.45

Source: Authors' own estimates based on primary data

Table 7.9 Net Income From Crop and Milk Production, Eastern Uttar Pradesh

Type of Pump	*Type of Farmer*	*Gross Cropped Area (Ha)*	*Net Income From Crop Production (Rs/annum)*	*Net Income From Milk Production (Rs/annum)*	*Farm-Level Net Income (Rs)*	*Farm-Level Net Income (Rs/Ha)*
Diesel Well	Well owner	5.29	124,587.3	7,152.3	131,739.6	24,880.2
	Water buyer	2.21	54,637.6	6,165.0	60,802.6	27,570.1
Electric Well	Water buyer	5.66	74,764.5	7,429.5	82,193.9	14,528.1

Source: Authors' own estimates based on primary data

Table 7.10 Net Income From Crop and Milk Production, South Bihar Plains

Type of Pump	*Type of Farmer*	*Gross Cropped Area (Ha)*	*Net Income From Crop Production (Rs/annum)*	*Net Income From Milk Production (Rs/annum)*	*Farm-Level Net Income (Crop + Milk) (Rs)*	*Farm-Level Net Income (Rs/Ha)*
Diesel Well	Well owner	3.14	11,1736.7	10,292.6	122,029.3	38,862.8
	Water buyer	1.70	61,517.7	8,130.89	69,648.6	40,969.8
Electric Well	Water buyer	2.49	140,105.5	9,958.09	150,063.6	60,266.5

Source: Authors' own estimates based on primary data

Findings and conclusions

The findings from the study can be summarized as follows. First, the rising cost of diesel has increased the cost of well irrigation for diesel well owners to an extent of 32% in south Bihar and 18% in eastern UP over the 16-year period from 1990–2006. But this did not have any positive impact on the price at which water is sold by diesel well owners due to the reducing monopoly power of diesel well owners over time. The actual price at which water is available to the water buyers reduced by 38% in south Bihar and 7.5% in eastern UP. Second, comparative analysis of irrigation water use, income from crops, dairying and entire farms; and water productivity in crop and milk production; and at the farm-level of three different categories of farmers, viz., water buyers in electric well commands, diesel well owners and water buyers in diesel well commands shows that higher cost of irrigation water motivates farmers to use irrigation water more efficiently from a physical point of view to minimize the cost of irrigation. Further, the farmers who are paying higher cost for irrigation water use it more efficiently also from agronomic and economic points of view, as reflected by higher values of water productivity in both physical and economic terms they obtain.

Third, the farmers who are paying a higher price for irrigation water use it more efficiently from economic point of view at the farm level than those who pay a lower cost, by optimizing crop and dairy inputs and allocating more area under crops yielding higher returns. This is reflected in the highest cropping system water productivity and farm-level water productivity in economic terms for the water buyers in diesel well commands. Fourth, the net income return farmers obtain from irrigated farming is not found to be elastic to the cost of irrigation water. The water buyers in diesel well commands get equal net returns per ha of land as the water buyers in electric well commands. They manage to sustain the net returns by minimizing the input costs and maximizing the returns, and by selecting crops that give higher returns per unit land. Fifth, the impact of diesel price on irrigation cost incurred by diesel well owners is not significant. Also, this burden is not passed on to the water buyers, owing to the lowering monopoly power of pump owners. Further, it was found that farmers would be able to cope with a steep rise in irrigation costs through irrigation-efficiency improvements and allocating more area under crops that give higher returns per unit of land and water. This in combination enhances the returns from every unit of water and energy used. By doing this, they are able to sustain the income from farming. This means that the rise in cost of diesel in real terms cannot make any negative impacts on economic prospects of diesel well irrigators, including water buyers.

The findings contradict the argument made by some scholars in the recent past (Mukherjee, 2007; Shah, 2007) by indicating that the impact of recent diesel price hikes on irrigation cost is not significant for diesel well owners. One reason for this is that the regions that are heavily dependent on diesel pumps for irrigation have a very shallow groundwater table. Also, over the past 2 decades, there has been an explosion in irrigation pump sets in eastern India (Kishore, 2004; Shah, 2001), and this drastically reduced the monopoly power of diesel pump owners. The monopoly rents charged by well owners to water buyers were so high, allowing them to absorb the slight shock caused by rising diesel prices on pumping costs, even as they were constrained in passing on rising cost to water buyers by growing competitive pressure.

We conclude that farmers would be able to cope with a high rise in irrigation costs through irrigation-efficiency improvements and allocating more area under crops that give higher returns per unit of land and that enhance the farming returns from every unit of water and energy used. This is because, when confronted with high irrigation costs, farmers would show greater willingness to invest labor and time to improve on-farm water application efficiency and take risks by adopting high-valued crops that are vulnerable to diseases and experience high price fluctuations in the market, such as potato and onion. This enables them to maintain almost the same net returns from farming as previously.

An important conclusion emerging from the study, which has significant relevance for public policy, is that raising the power (electricity) tariff in the farm sector to levels that the diesel pump owners pay today for energy use in irrigation will not cause any negative impact on the economic prospects of farming, provided the quality of electricity supply in rural areas is improved. Further, the study has great relevance for framing policies for mitigating the carbon emissions in developing economies of South Asia and Africa, where use of fossil fuel in agriculture is either already very high or is on the rise. The Indo-Gangetic basin in India has one of the largest carbon footprints because of the burning of fossil fuel for running around 4 million diesel engines.

Framing of the right kind of policies for pricing of diesel in the agriculture sector can bring about significant reductions in carbon dioxide emissions, thereby reducing the carbon footprint in the region's agriculture. We have estimated that the running of 6 million diesel pump sets in agriculture in the entire country would emit around 1.75 m. ton of carbon every year. This is based on the following assumptions: Each diesel pump set would run for an average of 1,000 hours every year, with an average suction depth of 20 feet and with a pump output of 20 liters of water per hour. The average energy consumption per pump set would be 4,000 kilowatt hour, which would require 400 liters of diesel per year,[2] and the burning of one liter of diesel would produce 0.732 kg of carbon (Nelson and Robertson, 2008).We do not recommend deregulation, which has already happened in India. Even if we improve the water and diesel use efficiency by 20%, the reduction in carbon emission would be 0.35 m. ton per year. Heavy subsidies for diesel used in the farm sector in shallow groundwater areas would also cause damage to agricultural economy and energy economy.

Notes

1. This chapter heavily draws from the paper titled 'Have Diesel Price Hikes Actually Led to Farmer Distress in India?' published in *Water International*, *35*(3), May 2010, pp. 270–284, authored by M. Dinesh Kumar, O. P. Singh and M. V. K. Sivamohan.
2. Using the formula that one liter of diesel had an energy content of approximately 10 kilowatt hours (Nelson and Robertson, 2008).

References

Byerlee, D., Diao, X. and Jackson, C. (2005). Agriculture, Rural Development, and Pro-Poor Growth Country Experiences in the Post Reform Era, Agriculture and Rural Development. Discussion Paper 21. Washington DC: The World Bank.

Datt, G. and Ravallion, M. (1998). Farm Productivity and Rural Poverty in India. *Journal of Development Studies*, *34*(4), 62–85.

Evenson, R. E., Pray, C. E. and Rosegrant, M. W. (1999). Agricultural Research and Productivity Growth in India. Research Report 109. International Food Policy Research Institute. Washington D.C., USA.

Kishore, A. (2004). Understanding Agrarian Impasse in Bihar. *Economic and Political Weekly*, *Review of Agriculture*, *XXXIX*(31), 3484–3491.

Kumar, M. D. (2003). *Food Security and Sustainable Agriculture in India: The Water Management Challenge*. IWMI Working Paper 60. Colombo: International Water Management Institute.

Kumar, M. D. (2007). *Groundwater Management in India: Physical, Institutional and Policy Alternatives*. New Delhi: Sage Publications.

Mukherjee, A. (2007). The Energy-Irrigation Nexus and Its Impact on Groundwater Markets in Eastern Indo-Gangetic Basin: Evidence From West Bengal, India. *Energy Policy*, *35*(12), 6413–6430.

Nelson, G. C. and Robertson, R. (2008).*Estimating the Contribution of Groundwater Irrigation Pumping to CO2 Emissions in India*. Washington, DC: International Food Policy Research Institute.

Ravallion, M. and Chen, S. (2004). *China's (Uneven) Progress Against Poverty.* Development Research Report. Washington, DC: The World Bank.

Shah, T. (2001). *Wells and Welfare in Ganga Basin: Public Policy and Private Initiative in Eastern Uttar Pradesh, India*. Research Report 54. Colombo, Sri Lanka: International Water Management Institute.

Shah, T. (2007). Crop per Drop of Diesel: Energy Squeeze on India's Small Holder Irrigation. *Economic and Political Weekly, 42*(9), 4002–4009.

Singh, O. P (2004) Water Intensity of North Gujarat Dairy Industry: Why Dairy Industry Should Take a Serious Look at Irrigation? Paper presented at the Second International Conference of the Asia Pacific Association of Hydrology and Water Resources (APHW 2004), Singapore, June 5–9, 2004.

8 Breaking the agrarian impasse in eastern India

M. Dinesh Kumar, Nitin Bassi, M. V. K. Sivamohan and L. Venkatachalam

Introduction

Eastern India has been at the center of the debate on agricultural growth for the stagnation it witnessed over the past 2 decades or so, since the miracle growth in West Bengal (WB) following land reforms, land consolidation and irrigation development. Among the four regions in the country, the total factor productivity (TFP) growth has been the lowest in the eastern region (Evenson et al., 1999). Scholars in agricultural economics and policymakers alike are rather clueless about the agrarian impasse in the states of WB, Bihar and Orissa, in lieu of the fact that the region encompassing these states has good endowment of natural resources, particularly water, a crucial input for agricultural growth.

A few researchers have argued that groundwater development and cheap well irrigation could trigger agricultural growth and economic prosperity in this groundwater-abundant region (Shah, 2001; Pant, 2004; Mukherji et al., 2012; Sharma, 2009; Sharma, 2011). Consequently, the public policy debates had centered around the impact of policies relating to irrigation on access equity in groundwater, particularly the economic impact of well irrigation on different classes of farmers in the region (Shah, 1993; Saleth, 1997; Kumar, 2007). The central question was appropriate policies for promoting well irrigation among the poor small and marginal farmers. While the public tube well program has failed in states like UP, the focus of researchers shifted to policy instruments for promoting efficient groundwater markets (Kumar, 2007). But such debates have paid little attention to the range of physical (climatic and ecological) and socio-economic factors (land-holding pattern and land fragmentation), which could severely limit agricultural prospects in a policy that is driven by the strategy of 'intensive use of water.'

Researchers argued that in eastern parts of India, rural de-electrification is taking place on a large scale and that it has a significant impact on well irrigation. They point out that with deteriorating quality of power supply in rural areas, farmers who were using electric motors to pump groundwater are now shifting to diesel engines (Kishore, 2004; Debroy and Shah, 2003; Mukherji, 2005). This, according to them, had affected not only the cost of abstraction of groundwater, but also influenced the functioning of groundwater/pump rental markets with rising charges for irrigation services. West Bengal has one of the lowest agricultural power connections, as compared to the agriculturally advanced states like Punjab, Haryana, Gujarat, Andhra Pradesh and Karnataka.

Some have advocated policy interventions like energy subsidies, subsidies for diesel pumps and low flat rates, for promoting equity in groundwater in eastern India (see, for instance, Kishore, 2004; Mukherji, 2003; Mukherji et al., 2012; Shah, 2001). The theoretical proposition is that the fuel charges or pump prices should be kept low, or good quality electricity be supplied and the same charged on a flat rate basis so as to keep: 1) the marginal cost of pumping very low in the first case, or 2) farmers' incentive for pumping more and more groundwater is high, thereby reducing the implicit cost in the second case. These were arguments made by researchers in the past (for instance, see Shah, 1993; Mukherji, 2007). This is also plausible for regions such as eastern India where groundwater exists in plenty. In eastern India, electricity policy would have some impact on irrigation water charge, as energy cost is a major component of the total cost of pumping water in shallow groundwater areas, provided a majority of well owners own electric pump sets. That said, how to deal with the issue of 'monopoly power' of well owners, which influences the selling price of water, has not been touched upon by the scholars working on public policies for promoting equitable access to groundwater (Kumar, 2007). The discussion, instead, revolved around the ways to reduce the cost incurred by well owners for pumping water.

In response to the deteriorating financial condition, the WB Power Distribution Company Ltd. introduced metering and pro rata pricing of electricity supplied to farmers in 2007. Conventional wisdom suggests that this improves both water and energy use efficiency in agriculture, while critics of this policy argue that it had resulted in shrinking water markets affecting the poor small and marginal farmers (Mukherji, 2008; Mukherji et al., 2012). The recent policy decision of the government of WB, to provide subsidized power connections to farmers and to relax the norms for issuing permits for well drilling, is lauded by some as a powerful policy instrument for bringing about a second Green Revolution in eastern Indian states. In this paper, we analyze the potential impacts of these policies on West Bengal's agriculture, its poor farmers who are dependent on water purchase and its groundwater resources.

What is the new policy?

The new policy of the WB government is based on the premise that in the groundwater-rich state, the poor farmers are deprived of their rightful access to groundwater for irrigation. It was said that the farmers in the state have been shying away from irrigated agriculture because of two important constraints induced by the wrong policies of the government in the past: the full cost that the state electricity board had been charging farmers for providing power connections for agricultural pump sets, which was estimated to be in the range of Rs. 50,000–Rs. 200,000 ($1,000–$4,000 USD); and the mandatory stipulation for farmers to obtain a license from the State Water Investigation Department (SWID) to apply for power connection for agro wells. Some scholars lament that the state electricity department was charging irrationally for providing power connections (IWMI, 2012; *Times of India*, 2012; Mukherji et al., 2012).

The recent policy change that the government of West Bengal brought in calls for 'rationalizing' charges for power connections and removal of restrictions on drilling of wells in the blocks that are categorized as 'safe' for groundwater exploitation. As per the new policy, the farmers need to pay only Rs. 1,000 to Rs. 30,000 for a new connection depending on the connected load. The rationalization here means that the farmers now have to pay for a new connection – which involves wires and electric poles and sometimes transformers – on the basis of connected load, and not on the basis of the length of the extension and the number of electric poles required (*Times of India*, 2012). Subsequently, there has been a lot of 'drum beating' from certain quarters of the academia and research fraternity that these changes are the outcome of their intensive field research and long-term engagement with policymakers from that region. Their main contention is that this would 'kick-start' a second Green Revolution in the state, and therefore other eastern Indian states should follow suit (Mukherji et al., 2012; IWMI, 2012).

Were there major policy constraints to agricultural growth in West Bengal?

It is well known that agriculture revolution swept West Bengal during the 1980s with the value of crop outputs jumping from 42 billion rupees to 75 billion rupees from 1980–83 to 1990–93, mainly as a result of land reforms undertaken in the state after 1977–78 and provision of adequate public supply of credit and other inputs to small farmers (Ramchandran et al., 2003). Between 1980–83 and 1990–93, the state recorded the highest annual compounded growth rate in crop yields (of 4.8%) in terms of rupees per ha of GCA (gross cropped area) and crop value (6%). Against this, the growth rate in crop yield was only 1.9% per annum during 1990–93 to 2000–03, and that in crop value was 2.4% per annum during the same period. Some attribute the recent stagnation in agricultural growth to lack of expansion in irrigation facilities (IIT, 2011).

It is argued that tube well irrigation in WB hasn't grown at the same pace as in some of the agriculturally prosperous states such as AP, Karnataka, Punjab and Haryana. This, according to Mukherji et al. (2012), was because of the wrong policy of the state's electricity board of charging the full cost of new power connections to the farmers and the corrupt practices of its groundwater department in issuing well-drilling licenses. They point out that the same has created a major bottleneck for agricultural growth in that state. First of all, it is absurd to compare a high rainfall sub-humid region like WB to low-to-medium rainfall semi-arid states like AP, Punjab and Haryana in assessing the importance of irrigation for agricultural growth.

The fact that marginal returns from irrigation would be much less in high rainfall, water-rich regions as compared to low rainfall, water-scarce regions is well documented. The farmers in high rainfall regions often take two crops without irrigation, whereas in low rainfall semi-arid to arid regions, even the monsoon crop might require 1–2 supplementary irrigations for maturity. Moreover, the aquifers in AP and Karnataka are hard rock in nature, with extremely limited groundwater potential, and therefore would require much higher well density to support the same extent of irrigation as WB.

Historical data on the number of new agricultural power connections in the state of WB were used to make the point that rent seeking and corruption in SWID is rampant (IWMI, 2012; Mukherji et al., 2012). It should be noted that the new agricultural power connections in WB peaked in 1989, and the number has been declining since then to become almost nil in 2003. But the blame is put conveniently on the state electricity company, i.e., WBSEDCL (West Bengal State Electricity Distribution Company Ltd.) and the groundwater department, by saying that the restrictive policies such as the Groundwater (Regulation and Control) Act of 2005, designed to control new wells and prepare an inventory of wells, and the high connection charges led to the electricity department virtually stopping new connections since 2003 (IWMI, 2012). While one wonders how policy changes brought in 2003 affect irrigation growth in the 1990s, the critics of the Act refuse to provide any data to see whether the number of applications from the farmers for new power connections and the connection charges has changed between 1989 and 2003.

One can see that the issue of corruption is overplayed. The only piece of information researchers furnish to highlight the issue of corruption is that till September 2010, a total of 23,000 applications were received by the SWID officials, of which 8,500 were approved (see Mukherji et al., 2012). This in no way makes the case of rent seeking or corruption, as applications for electricity connections can be rejected on several grounds, including the applicant's farm being located in one of the semi-critical blocks. This is a system followed in many Indian states for the past several years. In fact, a limited number of applications granted may also reflect that corruption is not all that rampant.

Even if a high rate of rural electrification is achieved, there are serious limitations to the extent to which it can contribute to increasing food grain production in eastern India. The constraints come from poor land availability (Kumar, 2003); very little additional land that can be brought under irrigation (GOI, 1999; Kumar, 2003); poor public investments in rural infrastructure including irrigation (NRAA, 2011); ecological constraints due to floods during the rainy season (Ladha et al., 2000); and overall lack of institutional and policy reforms in the agriculture sector (NRAA, 2011; Sharma, 2011). Small operational holdings and low crop yields prevent farmers from generating surplus. Hence, they are unable to invest in seeds of high-yielding varieties and the irrigation facilities required. Fragmentation of land forces farmers to depend on water sellers rather than investing in their own irrigation infrastructure, which would be economically inefficient owing to poor utilization of the potential (Kumar et al., 2012). The above analysis shows that there are major physical and ecological constraints to agricultural growth in West Bengal and the other eastern Indian States.

Claims of poverty reduction: Too big, too early

It is judged that the latest policy would lead to hundreds of thousands of farmers applying for electricity connections, with a resultant expansion in well-irrigated area. This, according to its proponents, would reduce poverty in West Bengal and other eastern states. It has been said that even if half a million farmers in WB take to electricity connections for pump sets, it would lead to an additional irrigated area of 3.7 million ha, and if 50% of this target is achieved, the additional area would be 1.85 million ha (IWMI, 2012). In the subsequent sections, we will see that even these calculations, based on basic statistics, are tailored and faulty.

While one needs to seriously question the logic behind making early predictions of major poverty-reduction impacts of such a policy shift, there are many fundamental issues that these policy prescriptions raise: 1) whether there is potential for groundwater irrigation in West Bengal for expansion; 2) whether the new policy decision of the state would actually lead to poor farmers benefiting from highly subsidized power connections, and be able to move them out of poverty trap; and 3) what would be the likely impact of the policy decision on the state's economy and social welfare.

Creating irrigation infrastructure without looking at demand

In order to address the questions raised above, we need to revisit a few facts about West Bengal's agriculture, which were conveniently goofed up. First, West Bengal is one of the most intensively cropped and intensively irrigated states in the country, with cropping intensity ranging from a lowest of 147% to a highest of 252% (see Table 8.1).

Against the net sown area of 5.25 million ha, the gross cropped area is 9.5 million ha. Against the net irrigated area of 3.11 million ha, the gross irrigation is to the tune of 5.5 million ha. The irrigation intensity is close to 176%. This means much of the irrigated land receives irrigation in two seasons. The state already has 5 lac plus agricultural tube wells, of which more than one lac are operated by electric pump sets, and the remaining by diesel engines.

That said, the amount of additional area that can be irrigated through new power connections to farms would be decided by three important factors. The first factor is the amount of cultivable land, which is in rain-fed due to lack of irrigation water. As the statistics presented earlier, the state has only 2.1 m. ha of land, which is cultivated but remains unirrigated for multiple reasons. The second factor is the agro climate in the region/area in question, mainly the rainfall and agro meteorological factors, viz., humidity, temperature and wind speed. Obviously, farmers would irrigate only if the rainfall cannot meet the crop water requirements.

Table 8.1 Cropping Intensity in Different Districts

Sr. No	*Name of District*	*Cultivable Area '000 Ha*	*Gross Cropped Area '000 Ha*	*Net Sown Area '000 Ha*	*Cropping Intensity (%)*
1	Darjeeling	157	194	132	147
2	Jalpaiguri	353	546	336	163
3	Coochbehar	258	521	251	207
4	Uttar Dinajpur	279	488	276	177
5	Dakshin Dinajpur	188	308	186	166
6	Malda	282	443	216	205
7	Murshidabad	100	932	398	234
8	Nadia	300	668	291	230
9	North 24 Parganas	263	472	220	215
10	South 24 Parganas	381	524	358	147
11	Howrah	87	157	80	195
12	Hooghly	218	540	214	252
13	Burdwan	467	865	454	190
14	Birbhum	336	529	322	164
15	Bankura	389	514	350	147
16	Purulia	438	335	319	105
17	Paschim Medinipur	594	940	564	167
18	Purba Medinipur	293	554	289	192
19	West Bengal	5,682	9,530	5,256	181

Source: Agricultural Statistics, West Bengal, 2009–10.

Interestingly, the rainfall in West Bengal is one of the highest in the country, with the mean annual rainfall varying from 1,500 mm to 2,700 mm. So, many areas won't require irrigation at all, and farmers would be able to take the second crop merely using the soil moisture from monsoon rains, as is evident from the high cropping intensity in most of the districts of the state. So, it is quite unlikely that the 2.1 m. ha of cultivated land would ever be available for irrigation. But it was mentioned that a total of 3.7 m. ha would be the additional irrigation potential from energized pump sets. Clearly, such a utopian idea calls for reclaiming and annexing land from the Bay of Bengal!

The third point, which calls for more detailed discussion and greater understanding of irrigation economics, is the viability of irrigation using pump sets for the poor small and marginal farmers, whom the new policy is expected to immensely benefit. It is again a fact that West Bengal has the largest percentage of marginal holdings amongst the major states in the country, with farmers having less than 1.0 ha of land, constituting 85% of the total farm holdings. It is logical to believe that all those well owners in the state, who already have electric and diesel pump sets, are large and medium farmers, and most of those who are left out are small and marginal farmers. The statistics show that the average area irrigated by an electric pump set in West Bengal is 14.5 ha.[1] The well-owning farmers, apart from irrigating their own land, provide irrigation services to several small and marginal farmers, but at prohibitive prices. The marginal farmers prefer such an arrangement over owning an energized well because the latter is economically unviable. Even if power connections are free, and the land requires irrigation desperately, these farmers would not find it easy to drill a tube well and install electric pump sets and meters.

The data on well and pump ownership across India underscores this argument. They show that the ownership of wells and pump sets is lowest for small and marginal farmers and highest among large farmers (see Figure 8.1 for wells and Figure 8.2 for pumpsets) (Kumar, 2007). Even with heavy subsidy for well drilling and purchase of pump sets, well irrigation largely remained in the domain of large farmers who have the wherewithal to invest in well drilling and installation of pump sets (Kumar, 2007).

Even if the small and marginal farmers decide to install electric pump sets, the only gain will be in the form of the difference between the price of purchased water in the market and the variable cost of pumping water using their own well. But against this, the total capital cost – of drilling a well, installing a pump set and connection charges – per unit area of land would work out to be very high for the small holdings. Hence, they would find it unaffordable. Under such circumstances, it is quite possible that the resource-rich diesel well owners would try and obtain the benefit of subsidized power connection. Nevertheless, the larger issue that the policy change raises is even more serious: What welfare gains does a state, which is on the brink of bankruptcy, make from investing a sum of Rs. 50,000–Rs. 100,000 for power connection per farmer in the form of subsidy,

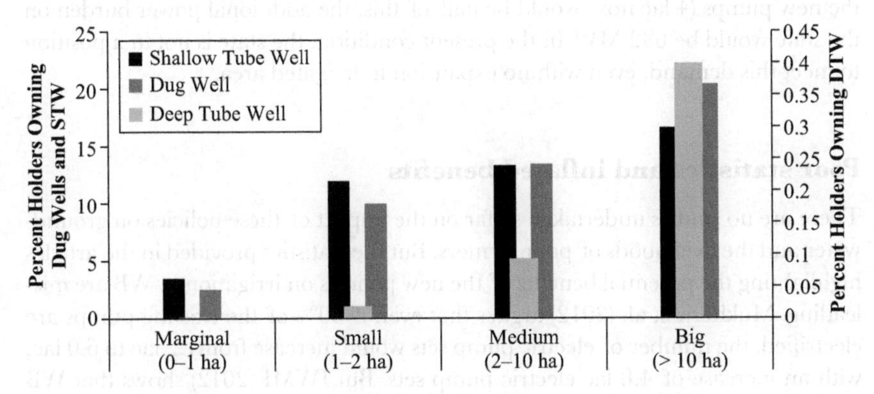

Figure 8.1 Ownership of Wells by Landholding Classes

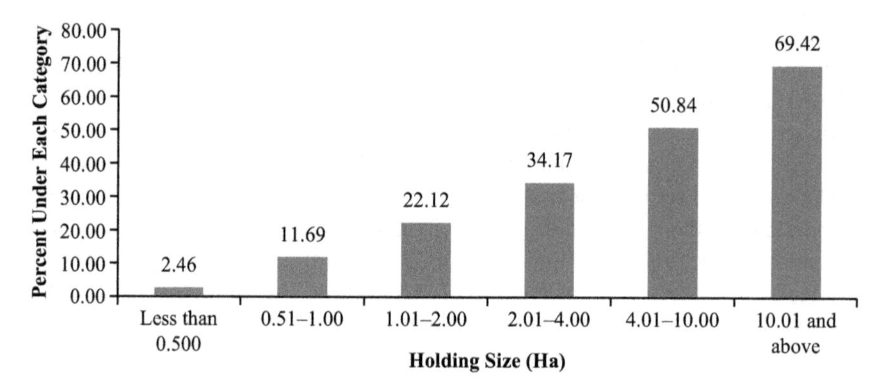

Figure 8.2 Skewed Ownership of Agricultural Pump Sets in India

when the intended beneficiary, who in this case is a small or marginal farmer, hardly gets any benefit? A much better policy prescription for the state would have been to subsidize micro diesel engines, which the small and marginal farmers would find not only economically viable, but also affordable. Here, the financial burden on the government would be much less.

Leaving aside the issue of welfare, when almost every Indian state is facing huge power shortages, it is hard to expect the government of West Bengal to handle the additional energy burden, which would be created by large farmers replacing their diesel engines with electric pumpsets to reduce their pumping costs. The state is already reeling under a severe power crisis. The average power shortage of the state is nearly 600 MW. The state's power utility, WBPDCL (West Bengal Power Development Corporation Ltd.) is finding it hard to make payments for the coal purchased from the subsidiaries of the Coal India Ltd., which has an outstanding due of Rs. 500 crore. Nearly 90 lac subscribers in West Bengal, barring Kolkata, now face daily power cuts as long as 4–5 hours (Business Line, 2011).

Today, the average power consumption by an agro well in WB is 6,493 units per annum. Even if we consider that the average annual electricity consumption by the new pumps (4 lac nos.) would be half of this, the additional power burden on the state would be 632 MW. In the present condition, the state is not in a position to meet this demand, even with no expansion in irrigated area.

Poor statistics and inflated benefits

There are no studies undertaken so far on the impact of these policies on groundwater and the livelihoods of poor farmers. But the statistics provided in the articles highlighting the potential benefits of the new policies on irrigation in WB are misleading. Mukherji et al. (2012) argues that even if 50% of the existing pumps are electrified, the number of electric pump sets would increase from1.2 lac to 6.0 lac, with an increase of 4.8 lac electric pump sets. But IWMI (2012) shows that WB has a total of only 5 lac pump sets (with over 1 lac electric pumps). Therefore, the maximum increase in electric pump sets possible through replacement of 50% of the existing pumps would only be 2.0 lac instead of 4.8 lac claimed by Mukherji et al. (2012). Clearly, use of such poor statistics does not support sound policymaking.

Attempts are made to inflate the economic benefits of such a policy shift. For instance, Mukherji et al. (2012) assume that an electric well irrigates nearly 7.9 ha of land, and hence estimate that the replacement of 4.8 lac diesel engines by electric pumps would lead to an added area of 3.7 m. ha (i.e., $7.9 \text{ X } 4.8 \times 10^5 = 3.7 \times 10^6$). Such a huge jump in irrigation would be possible only if we assume that the area previously irrigated by diesel well owners was zero. While it is logical to assume that diesel pump sets would have irrigated smaller area as compared to electric pump sets, mainly because their average land holdings are generally less (Kumar et al., 2010; Kumar et al., 2011), it would be totally incorrect to use the average area irrigated by electric pump sets as the basis for estimating the irrigated area expansion.

The estimates of the yield benefit are illustrative of this tendency. Mukherji et al. (2012) have assumed a yield gain of 2.5 ton/ha by replacing diesel engines with electric pumps, and thus predict that the incremental benefit would be in the tune of 4.62 m. ton. One is not sure about the source of this incremental yield when in the case of both diesel and electric well commands, the land is irrigated. On the contrary, research elsewhere shows that the diesel well owners often got higher yield when compared to their electric pump-owning counterparts, owing to greater control over irrigation water delivery (Institute of Rural Management Anand/UNICEF, 2001).

A much larger benefit to poor non-well-owning farmers of WB was anticipated through this new policy. It is argued that millions of poor farmers would participate in the water markets, with an increase in irrigation potential of existing tube wells through the replacement of diesel engines by electric pumps and lowering of the charges for pump rental service (Mukherji et al., 2012; IWMI, 2012). Such assumptions are faulty. First of all, the ability of the well-owning farmers to meet the demand for water from neighbors would be lower with electric engines, as they are dependent on the limited power supplied by the utility. Second, the benefit of reduced cost of pumping will not be transferred to the water buyers, and therefore the actual demand for water is not likely to increase with this new policy.

As pointed out earlier, in reality the actual gains from such a policy would be appropriated by the farmers who replace their old diesel engines because with electric power they would incur much lower cost for every unit volume of water pumped. This private benefit will come at a huge cost to the state exchequer. If we assume that the average cost to the State Electricity Department (in the form of subsidy for new power connection) for providing one connection would be Rs. 100,000, then the total cost of providing 4 lac new connections would be 4,000 crore rupees. For a state that is already in a state of bankruptcy, this is an enormous burden.

Policy flip flops

One of the landmark and bold decisions of the previous government in WB was to go for complete metering of farm wells and to charge for power supply on pro rata basis. As per the pricing policy, farmers who use electricity during day time (06.00–17.00 hours) would pay Rs. 1.55 per unit; those who use during 17.00 hours to 22.00 hours would pay Rs. 3.3 per unit; and those who use during 22.00 hours to 06.00 hours would pay only Rs. 0.97 per unit. Such a step has resulted in the reduction in revenue losses to the State Electricity Department. With this landmark policy initiative in this economically backward state of India, the long-held view that metering and raising power tariff in agriculture would find a lot of resistance from farming lobby and therefore would be politically unviable (Shah, 1993; Shah et al., 2004; Mukherji, 2008), stands questioned. According to GOI (2011), while the average cost of production and supply of electricity in WB was Rs. 4.83/kWh in 2011–12, the State Electricity Board recovered an average tariff of Rs. 2.69/kWh from agricultural consumers. Against this, the average tariff for agricultural consumers charged by respective state electricity utilities was only Rs. 0.32/kWh in Andhra Pradesh and almost zero in Tamil Nadu.

More importantly, as evidenced from other parts of India (Kumar et al., 2011), this pricing policy must have resulted in significant water and energy use efficiency in irrigated agriculture. Even the progressive states have been shying away from metering agricultural consumers, citing political economic compulsions. But the proponents of the new policy (of free power connection) vehemently argue that this is bad for water buyers. The best agricultural power pricing policy for West Bengal is to go for a judicious 'mix' of pro rata pricing and flat rate, claiming that such a mode of pricing would promote competitive water markets with lowering of water prices as well as water use efficiency. The IWMI report says:

> Metering of agricultural pump sets is already promoting water use efficiency, but a collateral damage has been a contraction in water markets, and hardest hit have been small holder farmers as they often rely on these markets for access to irrigation. (IWMI, 2012, p. 2)

The underlying argument is that when electricity is metered, the pump owners will not have any incentive to sell water, and even when they decide to sell water, they would pass on the cost burden induced by the marginal cost of pumping water to the buyers, and with that the price of water would go up (Mukherji, 2008). Field evidence hardly supports any of these arguments.

The diesel well owners, who incur positive marginal cost of pumping water, and who can be a proxy for electric pump owners who get metered power, were found to be selling water more aggressively to make profits than those who had electric submersible pumps and were paying for electricity on the basis of connected load. Against an average area of 22.8 ha irrigated by diesel well owners, the total area of water buyers was 19.2 ha – i.e., nearly 84% of the pumping was done to provide irrigation service to neighboring farmers – whereas in the case of electric well owners, against a total area of 27.0 ha irrigated by the tube well, the area irrigated by water buyers is 22.3 ha (82%). Hence, the argument that under metered tariff, well owners would have less incentive to sell water falls flat (Mukherji, 2008).

As regards the likely impact of metering on water prices, again one can compare the diesel well owners (as proxy for metered power supply) and electric pump owners under a flat rate system. But simple comparison of price of water would be unjustifiable as the costs incurred for pumping water vary widely between diesel well owners and electric well owners who pay flat rate charges. Therefore, it is important to look at the monopoly price ratios under the two situations.

Analysis of primary data from villages in WB (Mukherji, 2008) shows that under the flat rate system, which existed till 2007, the monopoly power enjoyed by the electric pump owners, estimated in terms of monopoly price ratio (MPR = 4628/1665 = 2.90) was much higher than that of diesel pump owners (MPR = 12998/7448 = 1.90). Again, these estimates seem to have considered a higher irrigation cost for well owners. Estimates using data on average irrigated area by electric pump owners show that the actual variable cost incurred by electric well owners was only Rs. 277/ha of irrigation under flat rate system[2] and not Rs. 1665/ha. But these owners charge Rs. 4628/ha for irrigation. This means that the monopoly price ratio for electric pump owner is 16.7 (i.e., 4628/277) and not 2.9. But the actual cost to the state electricity company of supplying power necessary for irrigating 1 ha of crop was worked out to be Rs. 2667/ha. This clearly shows that while the state provides a huge power subsidy to the well owner (Rs. 2667– Rs. 277 = Rs. 2390/ha), not even a fraction of it is passed on to the water buyers.

On the other hand, there is increasing empirical evidence now available from south Bihar and eastern UP to suggest that the price at which water is traded in the market or the pump rental charges is not decided by the cost of production of water, but the market conditions (Singh and Kumar, 2008). No matter what you do to reduce the cost incurred by well owners, the prices will not come down unless the monopoly power of the seller is reduced (Kumar, 2007; Singh and Kumar, 2008). Further, research showed that the mode of pricing of electricity does not influence the monopoly prices being charged by well owners in the market. But the flat rate pricing puts large well owners in a very advantageous position, as they could bring down their implicit unit cost of pumping groundwater. Therefore, pro rata pricing of electricity would promote equity in access to groundwater, if many farmers from the same area have access to electricity connections, reducing the monopoly power of well owners (Singh and Kumar, 2008).

Therefore, it is wrong to expect any major increase in price of water, even though under a metered tariff system the cost of pumping would go up from Rs. 277/ha to Rs. 1034/ha, as the monopoly power of the well owner does not change, and that only can change the market conditions.

That said, with the subsidized power connections, the profit margins of the old diesel well owners would go up substantially. While they would incur roughly Rs. 1,034/ha as the variable cost of pumping (against Rs. 7450/ha previously) with the new electric pump, the selling price of water would still be Rs. 13,000/ ha. This would be a windfall gain for them. For every ha of irrigation service provided, the income would be nearly Rs. 12,000, and therefore they would be able to recover all their capital investments for power connection and pump set in less than a year.

The potential impacts of the policy on wetlands

It is argued that abundant groundwater in WB and other parts of eastern India could be exploited to bring about agrarian change and poverty reduction (Shah, 2001; Mukherji, 2003, 2005, 2008; Mukherji et al., 2012; Sharma, 2009; Sikka and Bhatnagar, 2006) and that limiting exploitation of groundwater has a huge opportunity cost (Mukherji et al., 2012). However, due to limited amount of arable land and with much smaller proportion of it lying unirrigated, and the high rainfall, there are major limitations on the extent to which groundwater use for agriculture could be intensified in West Bengal. Yet further exploitation of groundwater in West Bengal should be done with utmost care. The reason is a small increase in groundwater pumping can play havoc with the region's water ecology, as thousands of wetlands receive their inflows also from shallow aquifers. Lowering of the water table of shallow aquifers during winter-summer seasons, when agricultural water demand actually picks up, can result in the temporary drying up of the shallow wetlands. This will have a huge impact on very poor families. The reason as pointed out by Mukherjee and Kumar (2012) is that the poor farmers depend on these water bodies not only for domestic water supplies and irrigation, but also for fish, which is a major source of protein. Clearly, the negative externalities of such approaches were ignored.

Arguments about recharging the aquifers during monsoons using schemes such as NREGS (National Rural Employment Guarantee Scheme) (IWMI, 2012) do not hold much water, as the aquifers are fully replenished and do not have much storage space during monsoon months. In a state that is endowed with large number of small water bodies (wetlands), which function as natural recharge systems, why should one invest so much money for exploiting groundwater and then recharging? Instead, if these wetlands are well managed, the poor farmers can tap water from these small water bodies for irrigation at much lower cost. The obsession with the idea of achieving 100% utilization of natural recharge can only be attributed to the poor understanding of how groundwater systems interact with wetlands. As is now understood, the concept of 'groundwater overdevelopment' is complex and is linked to various 'undesirable consequences,' which are physical, social, economic, ecological, environmental and ethical in nature (Custodio, 2000). Therefore, an assessment of groundwater overdevelopment involves hydrological, hydro-dynamic, economic, social and ethical considerations (Kumar and Singh, 2008; Kumar et al., 2012).

What can change the agrarian economy of eastern India?

Eastern India has a subtropical-to-tropical climate, and the region also receives high to very high rainfall. The future of farming in the region has to be based on the agro-ecology. The paddy-wheat system of cultivation has been very dominant in the eastern IGP for several decades (Pathak et al., 2003; Sikka and Bhatnagar, 2006), whereas paddy is the main crop in Orissa and Jharkhand (Sikka, 2009). Intensive irrigated paddy production, particularly cultivation of irrigated *boro* rice, had also resulted in the Green Revolution in West Bengal in the 1980s. But there are serious constraints in enhancing the income from this cropping system through yield improvement, induced by climate (Aggarwal et al., 2000; Pathak et al., 2003) and agro-ecology (Ladha et al., 2000). For instance, Pathak et al. (2003) found that the solar radiation during the growing season was much higher for the western side of the IGP (Ludhiana, Punjab) as compared to the eastern side (24-Parganas, WB) and minimum temperature much lower. Their study showed, using crop simulation models, that this difference in weather conditions could result in a difference in climatic potential yield between the two regions, to the extent of 3.0 t/ha (7.7t/ha in WB against 10.7t/ha in Punjab) in the case of rice and 2.7t/ha (5.2t/ha in WB against 7.9t/ha in Punjab) in the case of wheat.

The region needs farming systems that suit its agro-ecosystem. The region is characterized by large areas under wetlands, including areas that are under paddy grown under submerged conditions, areas that are inundated due to floods and tides, and numerous wetlands that have water year round. As regards the second point, there are large low-lying areas in the coastal region of West Bengal that are likely to get inundated during tides. They include coastal mangrove regions. In addition, there are floodplains that are likely to get flooded due to river flooding, mostly in north Bihar. These areas are suitable for extensive shrimp farming, with very little farm inputs. While the first category of areas, which get water from tidal exchange, would be suitable for saltwater shrimp, the second type of areas would be suitable for freshwater shrimp and numerous varieties of native fish, along with paddy.

In inland paddy-growing areas, which are in the very high rainfall region with shallow groundwater and alluvial plains, water from ponds and lakes could be used to irrigate the paddy fields, and shrimp can be farmed in the fields. The private ponds owned by farmers in their backyard and in the farms can be used for growing inland fish varieties such as rohu, catla and mrigal. These fish varieties grow to a marketable size of nearly 700 gram to 1 kilogram in a year. In groundwater-irrigated areas, the tanks can be replenished during the summer season using pumped groundwater. In the canal-irrigated areas of coastal Orissa, coastal plains of WB and Bihar (south and north), the ponds and tanks can be replenished by the release from canals.

The hard rock areas of eastern India, comprising western parts of WB, western Orissa and Jharkhand, have tropical, semi-arid climates with high rainfall. These region are not as water rich as the alluvial plains of West Bengal and north Bengal. A different type of farming system needs to be introduced. In these regions, though paddy is the dominant crop during kharif season, farmers grow many other crops, especially during winter season. In plots that are not traditionally used for raising paddy, farmers can take up vegetable production, and treadle pumps and micro diesel pumps can be used to irrigate these crops. Since the areas receive high rainfall (above 1,000 mm), small water-harvesting systems such as ponds can also be built, and used for supplementary irrigation of vegetables and fruits, along with fish rearing. The regions that are most ideal for this are the tribal areas of western Orissa; south Bihar and the western districts of West Bengal are also ideal for this.

Recent field-based research by ICAR (Indian Council of Agricultural Research) in Bihar and Orissa show that well-designed multiple-use systems can enhance the productivity of use of both land and water in eastern India remarkably. This involved integrating fisheries, prawn farming and duckeries with paddy irrigation using local secondary reservoirs for the water (Sikka, 2009).

Conclusions and the way forward

The recent policy decisions of the West Bengal government concerning farm power and groundwater sectors seems to be a 'knee-jerk' reaction to the realization that the agricultural growth in the state has almost stagnated since the 1990s after a quantum jump in outputs during the 1980s. The government has not taken cognizance of the macro level, physical and socio-economic constraints to agricultural growth such as limited arable land availability, an excessively large rural population depending on agriculture, huge power shortages facing the state and very small operational holdings of millions of farmers. Clearly, these constraints cannot be removed by offering heavily subsidized power connections in the farms and relaxing the norms for obtaining licenses for well drilling.

The argument that under metered tariff, water buyers are adversely hit with the rising price of water in the market lacks merit. The price at which water is traded in the market is decided by the overall market conditions, which have more to do with the monopoly power of well owners, and less to do with the cost of production of water. The current policy change can only lead to a situation of 'elite capture,' wherein many of the resource-rich big farmers would replace their diesel old engines with electric pumps and sell water to the marginal farmers, and make large profits at the cost of the public exchequer. In the long run, it could also spell doom for the myriad of wetlands, which support millions of poor rural families for their livelihoods, as the inflows into these freshwater bodies are sustained by outflows from shallow aquifers during the lean season.

While the state government must immediately take up a review of these policies, it must also formulate a scheme that offers micro diesel engines to small and marginal farmers at subsidized rates. It would not only be economically viable, but also financially feasible for the resource-poor farmers with very small holdings. The financial burden on the state also would be much less as compared to results from the new policy. But proper targeting of the subsidy would be essential to prevent pilferage and misappropriation by the rural elite. Simultaneously, managing the hundreds of thousands of small and large wetlands in the state through judicious management of groundwater would be crucial for improving the rural livelihoods and nutritional security of their people.

A policy intervention, which is based on a strategy for intensifying the use of land and water, will not work in a region where land use intensity is already very high. Therefore, a new policy for agricultural growth, which is driven by the strategy of enhancing the productivity of land and water and which is built on the concept of multiple use systems, needs to be adopted. Eastern India's agro-ecology is not homogeneous. Its rainfall varies from high rainfall to very high rainfall. It has hard rock areas and alluvial areas. It has semi-arid and sub-humid regions. It has low-lying areas that are flood prone to upland areas. The technical intervention for multiple-use systems needs to be designed, keeping in view the opportunities and the constraints the locality offers.

Notes

1. The average annual energy consumption by an electric pump set used in agriculture in West Bengal is 6,493 units. Using an average depth of pumping of 8 m and pump efficiency of 40%, the total volume of water pumped would be 1.69 lac m^3. Assuming an average delta of 1.2 m, this could result in an average irrigated area of 14.5 ha.
2. The average annual electricity charge paid by a submersible pump owner was only Rs. 5500, while that owner irrigated an average 20 ha of land. Here we do not consider the cost incurred by the well owner for power connection, submersible pump, well and motor, which are considered as sunk costs.

References

Aggarwal, P. K., Talukdar, K. K. and Mall, R. K. (2000). Potential Yields of the Rice-Wheat System in the Indo-Gangetic Plains of India. *Rice-Wheat Consortium Paper Series 10* (p. 11).New Delhi: Rice-Wheat Consortium for the Indo-Gangetic Plains, New Delhi, India, and Indian Agricultural Research Institute.

Business Line (2011, October 11). Power Crisis Looms in West Bengal as Utilities Run out of Cash for Coal.

Custodio, E. (2000). The Complex Concept of Over-Exploited Aquifer (2nd ed.). Madrid: Uso Intensivo de Las Agua Subterráneas.

Debroy, D.A. and Shah, T. (2003). Socio-Ecology of Groundwater Irrigation in India. In R. Llamas and E. Custodio (Eds.), *Intensive Use of Groundwater: Challenges and Opportunities* (pp. 1–88). The Netherlands: Balkema Publishers.

Evenson, R. E., Pray, C. E. and Rosegrant, M. W. (1999). *Agricultural Research and Productivity Growth in India.* Research Report 109. International Food Policy Research Institute: Washington, DC.

Government of India (1999). *Ninth Five Year Plan.* New Delhi: Author.

Government of India (2011, October). *Annual Report (2011–12) of the Working of the State Power Utilities and Electricity Departments.* New Delhi: Planning Commission, Government of India.

Indian Institute of Technology, Kanpur (2011, December). *Trends in Agriculture and Agricultural Practices in Ganga Basin: An Overview.* Report submitted to the Ministry of Environment and Forests, Indian Institute of Technology, Kanpur.

Institute of Rural Management, Anand/UNICEF (2001). *White Paper on Water in Gujarat.* Report submitted to the Department of Narmada Water Supplies, Government of Gujarat, Gandhinagar. Anand: IRMA.

International Water Management Institute (2012, April). *Agricultural Water Management Learning and Discussion Brief.* Colombo, Sri Lanka: AGWAT Solutions, improved livelihood for small holder farmers.

Kishore, A. (2004). Understanding Agrarian Impasse in Bihar. *Economic and Political Weekly, XXXIX*(31), 3484–3491.

Kumar, M.D. (2003). *Food Security and Sustainable Agriculture in India: The Water Management Challenge.* Working Paper #60. Colombo: Sri Lanka: International Water Management Institute.

Kumar, M. D. (2007). *Groundwater Management in India: Physical, Institutional and Policy Alternatives*. New Delhi: Sage Publications.

Kumar, M. D., Narayanamoorthy, A. and Sivamohan, M. V. K. (2010). *Pampered Views and Parrot Talks: In the Cause of Well Irrigation in India*. Occasional Paper #1. Hyderabad, India: Institute for Resource Analysis and Policy.

Kumar, M. D., Scott, C. and Singh, O. P. (2011). Inducing the Shift from Flat-Rate or Free Agricultural Power to Metered Supply: Implications for Groundwater Depletion and Power Sector Viability in India. *Journal of Hydrology, 409*(1–2), 382–394.

Kumar, M. D. and Singh, O. P. (2008). How Serious Are Groundwater Over-Exploitation Problems in India? A Fresh Investigation Into an Old Issue. In M. D. Kumar (Ed.), *Managing Water in the Face of Growing Scarcity, Inequity and Declining Returns: Exploring Fresh Approaches*. 7th Annual Partners' meet of IWMI-Tata Water Policy Research Program, ICRISAT, Patancheru, AP, April 2–4.

Kumar, M. D., Sivamohan, M. V. K. and Narayanamoorthy, A. (2012). The Food Security Challenge of the Food-Land-Water Nexus in India. *Food Security, 4*(4), 539–556.

Ladha, J. K., Fischer, K. S., Hossain, M., Hobbs, P.R. and Hardy, B. (Eds.). (2000). *Improving the Productivity and Sustainability of Rice-Wheat Systems of the Indo-Gangetic Plains: A Synthesis of NARS-IRRI Partnership Research*. Discussion Paper 40. Manila, Philippines: International Rice Research Institute.

Mukherjee, S. and Kumar, M.D. (2012). Economic Valuation of a Multiple Use Wetland Water System: A Case Study From India. *Water Policy, 14*(1), 80–98.

Mukherji, A. (2003). Groundwater Development and Agrarian Change in Eastern India, IWMI-Tata Comment #9. Based on Ballabh, V., Chaudhary, K., Pandey, S. and Mishra, S. IWMI-Tata Water Policy Research Program.

Mukherji, A. (2005). Is Intensive Use of Groundwater a Solution to World's Water Crisis? In *Water Crisis: Myth or Reality? Selected Papers From a Workshop on the Same Issue*. Organized by Harvard University and Marcelino Botin Foundation, Santander, Spain, June 14–16, 2004.

Mukherji, A. (2007). The Energy-Irrigation Nexus and Its Impact on Groundwater Markets in Eastern Indo-Gangetic Basin: Evidence From West Bengal, India. *Energy Policy, 35*(12), 6413–6430.

Mukherji, A. (2008). *The Paradox of Groundwater Scarcity Amidst Plenty and Its Implications for Food Security and Poverty Alleviation in West Bengal, India: What Can Be Done to Ameliorate the Crisis?* Paper presented at 9th Annual Global Development Network Conference, Brisbane, Australia, January 29–31.

Mukherji, A., Shah, T. and Banerjee, P. (2012). Kick-Starting a Second Green Revolution in Bengal, Commentary. *Economic and Political Weekly, 47*(18), 27–30.

National Rain-Fed Area Authority (2011). Challenges of Food Security and Its Management. New Delhi: Author.

Pant, N. (2004). Trends in Groundwater Irrigation in Eastern and Western UP. *Economic and Political Weekly, XXXIX*(31), 3463–3467.

Pathak, H., Ladha, J. K., Aggarwal, P. K., Peng, S., Das, S., Singh, Y., . . . Gupta, R. K. (2003). Trends of Climatic Potential and On-Farm Yields of Rice and Wheat in the Indo-Gangetic Plains. *Fields Crops Research, 80*(3), 223–234.

Ramchandran, V. K., Swaminathan, M. and Rawal, V. (2003). *Agricultural Growth in West Bengal*. Paper presented in the All India Conference on Agriculture and Rural Society in Contemporary India, Barddhman, December 17–20.

Saleth, R. M. (1997). Power Tariff Policy for Groundwater Regulation: Efficiency, Equity and Sustainability. *Artha Vijnana, 39*(3), 312–322.

Shah, T. (1993). *Groundwater Markets and Irrigation Development in India: Political Economy and Practical Policy*. New Delhi: Oxford University Press.

Shah, T. (2001). *Wells and Welfare in Ganga Basin: Public Policy and Private Initiative in Eastern Uttar Pradesh, India*. Research Report 54. Colombo, Sri Lanka: International Water Management Institute.

Shah, T., Scott, C. A., Kishore, A. and Sharma, A. (2004). *Energy Irrigation Nexus in South Asia: Improving Groundwater Conservation and Power Sector Viability*. Research Report No. 70. Colombo, Sri Lanka: International Water Management Institute.

Sharma, K. D. (2009). Groundwater Management for Food Security in India, Opinion. *Current Science, 96*(11), 1444–1447.

Sharma, K. D. (2011). Rain-Fed Agriculture Could Meet the Challenges of Food Security in India, Opinion. *Current Science, 100*(11), 1615–1616.

Sikka, A. K. and Bhatnagar, P.R. (2006). Realizing the Potential: Using Pumps to Enhance Productivity in the Eastern Indo-Gangetic Plains. In B. R. Sharma, K. G. Villholthand and K. D. Sharma (Eds.), *Groundwater Research and Management: Integrating Science Into Management Decisions*. Groundwater Governance in Asia Series-1. Colombo, Sri Lanka: IWMI.

Sikka, A. K. (2009). Water Productivity of Different Agricultural Systems. In M. D. Kumar and U. Amarasinghe (Eds.), *Water Sector Perspective Plan for India: Potential Contributions From Water Productivity Improvements* (pp. 73–84). Draft report. Colombo: International Water Management Institute.

Singh, O. P. and Kumar, M. D. (2008). Electricity Pricing be a Tool for Efficiency, Equity and Sustainability in Groundwater Use. In M. D. Kumar (Ed.), *Managing Water in the Face of Growing Scarcity, Inequity and Declining Returns: Exploring Fresh Approaches*. 7th Annual Partners' Meet of IWMI-Tata Water Policy Research Program, ICRISAT, Patancheru, AP, April 2–4.

Times of India (2012, September 2). Groundwater Irrigation Actually Reduces Poverty, Interview With A. Mukherjee.

9 Developing a household-level MUWS vulnerability index for rural areas

V. Niranjan, M. Dinesh Kumar and Yusuf Kabir

Water supply surveillance

Water supply surveillance is defined as 'the continuous and vigilant public health assessment and oversight of the safety and acceptability of water supplies' (WHO, 1976, 1993, 2004). Many millions of people, in particular throughout the developing world, use unreliable water supplies of poor quality, which are costly and are distant from their home (WHO and UNICEF, 2000). Over the years, there has been a growing realization that communities in the rural areas need water for productive as well as domestic uses, indicating the need for an increase in the quantity of the water supplied from public systems, along with quality (Renwick, 2008; Nicole, 2000; van Koppen et al., 2006). This is important for meeting the millennium development goals (van Koppen et al., 2006).

Traditionally, water supply surveillance generates data on the safety and adequacy of the drinking water supply in order to contribute to the protection of human health. Most current models of water supply surveillance come from developed countries and have significant shortcomings if directly applied in a developing-country context. Not only the socio-economic conditions, but also the nature of water supply services is different. Water supply services in developing countries often comprise a complex mixture of formal and informal services for both the 'served' and 'unserved' (Howard, 2005).

Many millions of households in India do not have access to 'tap' connections at home. Only 24.2% of the rural population have access to tap connections (Census of India, 2001), and as a result a majority of the rural population depends extensively on private wells, hand pumps, bore wells and ponds and tanks, which provide untreated water, for domestic water supply (NSSO, 1999), a trend also found in many other parts of the developing world (Gelinas et al., 1996; Rahman et al., 1997; Howard and Luyima, 1999). Given the informal nature of the sources and 'services,' the data on actual water use by the households by the communities are absent. The problem is compounded by the lack of clarity on the supply norms for fulfilling multiple water needs of rural population.

Nevertheless, the sources that are reliable and that can provide adequate quantities of water of sufficient quality to meet various productive and domestic needs seem to be far less than adequate. It is evident from the fact that the rural poor tend to compromise on their basic needs, with resultant undesirable outcomes on health and hygiene and livelihoods of rural communities. Therefore, a well-designed and well-implemented water supply surveillance in relation to domestic and productive needs of the community is important to provide input into water supply improvements. The key to designing such a program is information about the adequacy of water supplies and the health and livelihood security risks faced by populations due to lack of it at national or subnational levels. This will help identify areas that are vulnerable. But, as Nicole (2000) notes, there are a range of natural, physical, social, human, economic, financial and institutional factors influencing the vulnerability of the rural population to problems associated with inadequate supply of water for consumption and production needs. They are not captured in the traditional surveillance programs.

Past approaches to water supply surveillance

The inextricable link between water security, health, livelihood and economic gain is quite well established (Botkosal, 2009; HDR, 2006; Nicole, 2000). Improving water security of the poor brings about significant health and poverty reduction benefits (DFID, 2001; HDR, 2006; WHO, 2002). The economic losses due to deficit in water supply of sufficient quantity, quality and reliability are disproportionately higher for the poor communities. This is owing to greater risk of employment loss, health costs, loss of productive workforce and water-based livelihoods (HDR, 2006).

As Nicole (2000) argues, a demand-responsive approach to water supply requires that the livelihood needs of the community are also taken into account, rather than only the supply requirements for human consumption and sanitation needs. Therefore, an assessment of water supply at the household level, based on the old norms worked out on the notion of water supplies that serve human health and hygiene needs would be grossly inappropriate. In India, the monitoring of rural water supply is based on simplistic considerations, involving data on the number of households covered by different types of water supply systems and the characteristics of the sources. The data gathered through such surveys are silent on the amount of water actually consumed by the population, and the quality and reliability of the supplied water, all of which determine the health and livelihood outcomes.

Why vulnerability index for MUWS?

The foregoing discussion suggests that comprehensive approaches to water supply surveillance were, by and large, lacking for quite some time. The approaches to water supply surveillance that allow targeting of surveillance activities on vulnerable groups were assessed by Howard (2005), using case studies from Peru and Uganda. The Peru case study attempted to incorporate some measures of vulnerability into the surveillance program design through a process of 'zoning' that was based on water service characteristics, whereas the Uganda case study involved development of a semi-quantitative measures of community vulnerability to water-related diseases, to zone the urban areas and plan surveillance activities. The zoning used a categorization matrix, which was developed incorporating a quantitative measure of socio-economic status (education, sources of livelihood, family size and type of housing), population density and a composite measure of water availability and use (Howard, 2005).

But the main limitation of these approaches is that they try to assess the vulnerability of the household against lack of water for human consumption and sanitation. They do not take into account the multiple water needs of the community, particularly the poor in rural areas. There are many factors such as the family occupations, social profile and financial stability that determine the household water needs for productive purposes.

Identifying the most vulnerable groups is not an easy task due to the complex interplay of a wide range of factors. Factors such as poor reliability (continuity of supply), costs (affordability) and distance between a water source and the home may all lead households to depend on less safe sources, to reduce the volume of water used for hygiene purposes and to reduce spending on other essential goods, such as food (Lloyd and Bartram, 1991; Cairncross and Kinnear, 1992; Howard, 2005). The evidence suggests that water interventions targeted at poor populations provide significant health benefits and contribute to poverty alleviation (DFID, 2001; HDR, 2006; WHO, 2002). Though it appears that poverty is a major factor deciding vulnerability, it is just one of the many complex factors that would eventually determine the outcomes of a family's high vulnerability to lack of water for multiple uses.

The factors that can influence vulnerability of a household to problems associated with lack of water for multiple uses could be: 1) degree of access to water supplies for human consumption, personal hygiene and productive uses such as livestock consumption in terms of quantity and desired quality, and the level of use; 2) social profile and family occupations; 3) social institutions and ingenuity; 4) condition of water resources; 5) climatic factors; and 6) financial condition (Lloyd and Bartram, 1991; Cairncross and Kinnear, 1992; DFID, 2001; Howard, 2002; Hunter, 2003; Nicole, 2000; Sullivan, 2002; WHO, 2002). The second and fifth factors influence the vulnerability by changing the household water demand. This may not be always in terms of the quantum of water, but in terms of the reliability of the supply. The third and fourth factors can change the external environment, which influences water supply. Here again, the degree of access depends on the presence/absence of social institutions and local custom and traditions, which are quite characteristic of poor and developing countries.

Now climate has a major bearing on the adverse effect of lack of water for hygiene and environmental sanitation. In arid and semi-arid climates, breeding of water-related insect vectors would be less during hot weather conditions. In flood-prone areas and areas receiving high rainfall, the occurrence of water-based diseases is likely to be higher, and therefore more caution needs to be exercised in the disposal of human and animal excreta (Hunter, 2003). At the same time, the demand of water for meeting livestock needs, and irrigating fruit trees and kitchen gardens, etc., would increase with increase in aridity and temperature. The same applies to the demand for water for washing and bathing. Arid areas are also drought prone. Hence, there is a need to develop a composite index, which takes into account these complex factors in assessing the vulnerability of rural households to inadequate supplies of water to meet multiple needs so as to make surveillance more targeted.

Deriving a MUWS vulnerability index

We begin with the premise built on the knowledge from extensive review of past research studies dealing with related topics that the vulnerability of a household to inadequate supply of water to meet drinking water, sanitation and livelihoods needs is determined by four broad parameters: 1) capital assets and goods, 2) sequencing and time, 3) institutional linkage and 4) knowledge environment. The capital assets can be further divided into natural capital, social capital, physical capital and financial capital (Nicole, 2000). It is evident that while some of the capital asset-related parameters (physical capital and human capital) would determine the access to water supply and its use, the natural capital–related parameters, institutional linkage and knowledge environment would change the external environment that influences the supply and use for water. On the other hand, the capital assets such as natural capital, social capital and financial capital influence the demand for water.

All these parameters are factored in six broad sub-indices we have discussed previously. They are: 1) water supply and use, 2) family occupation and social profile, 3) presence of social institutions and ingenuity, 4) water resource endowment, 5) climate and drought proneness and 6) financial stability. Each one of these six broad factors constitutes one sub-index. The number of 'minor' factors, which together are considered to be influencing the measure of these sub-indices, the methods and procedure for their computation and sources of data are explained in Table 9.1.

The composite index of 'MUWS vulnerability' will have a maximum value of 10.0, meaning zero vulnerability; lower values of the index mean higher vulnerability. It is composed of six sub-indices (from A to F: in Table 9.1); each one will have unequal weightage in deciding the value of the index. The maximum value of sub-index A will be 3.0; that of B, C and D will be 1.0; and that of E and F will be 2.0 each. The sub-sub-index also will have equal weightage (measured on a scale of 0 to 1.0). The sum of the values of all sub-indices under sub-index A would be multiplied by 0.30 to obtain the value to be imputed into the mathematical formulation for estimating the composite index. The sum of the values of all sub-indices under sub-index B and D would be divided by 2 to obtain the value to be imputed into the mathematical formulation for estimating the composite index. The sum of sub-sub-indices under sub-index E would be multiplied by 0.50.

Table 9.1 Deriving a Household-Level MUWS Vulnerability Index

Sr. No	*Parameters*	*Quantitative Criterion for Measurement*	*Method of Data Collection*
A	**Water Supply and Use**		
1	Access to water supply source (primary)	Vulnerability decreases with improved access. Access is an inverse function of the distance. The index is a function of the distance to the source from 0 within the dwelling to a maximum of 1 km and above in gradations of 0.20.[1]	Primary survey
2	Frequency of water supplies	Vulnerability increases with decrease in frequency of water delivery.[2]	Ditto
3	Ownership of alternative water sources	Ownership of alternative water sources would increase the overall access and reduce the vulnerability.[3]	Ditto
4	Distance to the alternative source 'owned'	Distance to the alternative source would increase the vulnerability. Often, the alternative sources are farm wells, which are located outside the village.[4]	Ditto
5	Access to other alternative sources	Vulnerability decreases with number of alternative sources.[5]	Ditto
6	Capacity of domestic storage systems	Vulnerability to lack of regular water supplies decreases with an increase in volume of storage systems in place.[6]	Ditto
7	Quantity of water used	Vulnerability increases with a decrease in quantum of water used against the requirement. The vulnerability can be treated as zero when all the requirements in the household are fully met.[7]	Ditto
8	Quality (chemical, physical and bacteriological) of domestic water supplies	Poor quality of drinking water increases vulnerability; bacteriologically, physically and chemically pure is the best water.[8]	Lab test results/ perceptions
9	Total monthly water bill as a percentage of monthly income	Vulnerability increases with increasing percentage of total family income spent on water. An expenditure level of 10% of monthly income is treated as highest and most vulnerable.[9]	Primary survey

(Continued)

Table 9.1 (Continued)

Sr. No	*Parameters*	*Quantitative Criterion for Measurement*	*Method of Data Collection*
B	**Family Occupation and Social Profile**		
1	Family occupation	Vulnerability will be low for families having regular sources of livelihood that are not dependent on water. Those who are dependent on irrigated crop production are considered to be not vulnerable. But those who are dependent on dairying are vulnerable. The vulnerability will reduce if they depend on wage labor and other sources of livelihood that do not require water.[10]	Ditto
2	Social profile	Vulnerability is also a function of the social profile. The families having school-going children are more vulnerable to inadequate quantity, quality and reliability of water supplies. This is also the case with families having office-going adults. But the vulnerability would reduce with the presence of surplus labor availability.[11]	Ditto
C	**Social Institutions and Ingenuity**	Communities' vulnerability to the problems associated with lack of water increases in the absence of social/community institutions; social ingenuity.[12]	Primary survey (but qualitative to be obtained from discussions)
D	**Climate and Drought Proneness**		
1	Climate(whether semi-arid/arid/hyper-arid or sub-humid/humid)	Vulnerability to lack of water for environmental sanitation is a function of climate. It increases from hot and arid to hot and semi-arid to hot and sub-humid to hot and humid to cold and humid.[13]	Secondary data on climate
2	Aridity and drought proneness	Vulnerability due to lack of water for domestic uses and livestock increases with increase in aridity, as it would increase the demand for water for washing, bathing, livestock drinking and irrigation of vegetables and fruit trees. Aridity areas are also drought prone.[14]	

(Continued)

Table 9.1 (Continued)

Sr. No	*Parameters*	*Quantitative Criterion for Measurement*	*Method of Data Collection*
E	**Condition of Water Resources**		Ditto
1	Surface and groundwater availability in the area	A renewable water availability of 1,700 m^3 per capita per annum is considered adequate for a region or town, estimated at the level of river basin in which it is falling.[15]	Secondary data
2	Variability in resource condition	The higher the variability, the greater will be vulnerability.[16]	Ditto
3	Seasonal variation	Regions that experience high seasonal variation in water availability are highly vulnerable.[17]	Ditto
4	Vulnerability of the resource to pollution or contamination	Surface water is more vulnerable to pollution than groundwater. Shallow aquifers are more vulnerable than deep confined aquifers.[18]	
F	**Financial Stability**	Overall financial stability of the family would influence the vulnerability. This is different from the earnings from current occupations, such as savings in banks/post office and ownership of productive land that is not mortgaged.[19]	Primary survey

Notes

1. Within the dwelling is 1.0; within the premise is 0.80; within 0.2 km distance is 0.60; between 0.2 and 0.5 km is 0.4; 0.5 and 1.0 is 0.2 and more than 1.0 km is 0.
2. Frequency can be indexed as total hours of water supply in a week as a fraction of number of hours.
3. It is assumed that the ability to manage water would be highest in the case of a functional open well, followed by bore well, hand pump and farm pond in decreasing order. The value of the sub-index would be 1.0 in the case of ownership of a functional open well, followed by 0.70 for a bore well, 0.50 for ownership of a hand pump and 0.30 for ownership of a farm pond.
4. Within the premises is 1.0; within 0.2 km distance is 0.80; between 0.2 and 0.5 km is 0.6; 0.5 and 1.0 is 0.4 and more than 1.0 km is 0.20.
5. The value of sub-index for this attribute would be 1.0 if there are four alternate sources or more, and the value would decrease proportionately with a decrease in the number of alternative sources.
6. It would decrease with increase in the ratio of the actual storage capacity available to the storage capacity required, and the value of the index would be higher. The storage capacity required would be an inverse function of the frequency of water supply. If supply comes once daily but during odd hours, then it can be assumed that the volume of water for the entire day's use would be required to be stored. So, the storage capacity would be $n \times f$. If it comes during the daytime for less than an hour, then half the daily water use would be the storage requirement. For more than one hour, the storage requirement would be minimal (around 20 liters per capita). With alternate day water supply, it could be the $2 \times n \times f$. For once in 3 days, it would be $3 \times n \times f$ and likewise. For round-the-clock water supply, the storage requirement would be zero, and here the ratio can be assumed as 1.

7. This sub-index is computed by taking the volume of water used (x) as a fraction of the minimum required (n), i.e., $\frac{x}{n}$, where n = water requirement as per norms. The value of n should be estimated by considering the human requirement of 50 lpcd (basic survival need as suggested by Glieck, 1997); the animal requirements should be decided by the types of livestock and the size and the requirement for kitchen gardens.
8. The value of the sub-index m will be 0.33 if the water is pure either bacteriologically, physically or chemically. The value will be 1.0 if pure on all counts.
9. The value of the sub-index would be zero if the family spends 10% or more of its monthly income on obtaining domestic water supplies, and would keep on increasing with reducing amount of money spent on water bills. The mathematical formulation for computing the index therefore is [1 – WC/MI], where WC is the monthly expenditure on securing water supplies, and MI is the monthly family income.
10. The vulnerability induced by family occupation is considered to be zero if the adults in the family are engaged in livelihoods that are not dependent on water in the village. The vulnerability is also considered to be zero for families having crop production with their own irrigation facilities. The families purely dependent on dairy farming would be assumed to have highest vulnerability (1.0), as the water for cattle drinking will have to be managed.
11. For families having school-going children and office-going adults, the situation could be treated as most unfavorable. Here, the sub-index could be assumed as zero (lowest), meaning the highest vulnerability. For families having either one of these, the value could be assumed as 0.33. For families having neither of these, the value would be treated as 0.67. For families, having surplus labor in the household for fetching water from distance, the sub-index could be assumed as 1.0.
12. The value can range from zero for the absence of social institutions or ingenuity to 0.50 for presence of either of these, to 1.0 for the presence of both. Social institutions would include WATSAN committees (Y = 0.50; N = 0). Social ingenuity would include existence of water-sharing traditions between households during crisis (Y = 0.25; N = 0.0) and practice of reusing water in households – using bathing/washing water for toilet flushing, use of sand and ash for cleaning utensils, etc. (Y = 0.25; N = 0.0).
13. The value ranges from zero for cold and humid to 1.0 for hot and arid in increments of 0.20.
14. The value ranges from 1.0 for cold and humid to zero for hot and arid with reduction of 0.20.
15. The value of the index is computed by taking the amount of renewable water as a fraction of the desirable level of 1,700m^3.
16. The index is computed as an inverse function of the coefficient of variation in the rainfall variability in that basin/sub-basin ($1 - x/100$), where x is the coefficient of variation in rainfall.
17. For alluvial areas, the value of this index is considered as 1. For hard rocks, the value is considered as 0.3. For sedimentary and alluvial deposits, the value is considered as 0.65.
18. Shallow groundwater areas and river/stream/reservoirs in the vicinity of industries are highly vulnerable, with a value of the sub-index equal to zero; distant reservoirs in the remote virgin catchments and groundwater from deep confined aquifers have a pollution vulnerability index of 1.0; shallow groundwater in rural areas have medium vulnerability, with a value of 0.50.
19. A family having 1.0 ha of productive land per member in a semi-arid, water-scarce region, or 0.5 ha of productive land per member in a water-rich area, is considered to be financially stable, with zero vulnerability, and the vulnerability is assumed to increase gradually with reducing size of land owned, with highest vulnerability for the landless. Again, the lack of ownership of land can be compensated by income savings, with a total saving of Rs. 20,000 equivalent to 0.5 ha in a water-rich area and 1.0 ha in a water-scarce area. This index could be computed as $x/0.50$ for water-rich areas, and $x/1.0$ for water-scarce areas (where x is the land owned in ha).

Computation of MUWS vulnerability index for households in three Maharashtra villages

A preliminary survey was carried out in six villages from three locations in Maharashtra. The three locations for project were selected in such a way that they represent three distinct typologies. They are as follows: 1) a village located at the foot of the hilly region, characterized by high rainfall and plenty of local streams flowing down from high altitudes fed by base flows from hilly aquifers; 2) a village located in hard rock plateau areas with low to medium rainfall, with the rural water supplies heavily dependent on the limited groundwater resources in the Deccan trap formations; and 3) a village located in the foot of hilly forested land, falling in the assured rainfall zone, with extremely limited groundwater, but that has local streams.

From the six villages, three were selected for the action research. The following physical and socio-economic criteria were used in the selection process: 1) a maximum number of households in the village have access to domestic water supply through tap connections, meaning the physical access to water for drinking and cooking and other household uses is very good; 2) the villages are predominantly agrarian, but a significant section of the farm households are not able to meet their farming needs from the available water sources (ponds, tanks and wells), and therefore demand water for multiple purposes, including water for livestock and water for kitchen gardens to improve their livelihoods, from the public systems; and 3) the current public water supplies across seasons are less than adequate to meet these needs largely due to competition from agriculture, but conditions are favorable for augmenting the available supplies, through technological and institutional measures.

The selected villages are: Varoshi in Jawali taluka of Satara, Kerkatta in Latur taluka of Latur district and Chikhali in Jivati taluka of Chandrapur district. They represent three different agro-ecological zones in the state: one from the high rainfall zone of western Maharashtra located in the foothills of Western Ghats, with high humidity; another from the low-to-medium rainfall zone of central plateau, which is drought prone and experiencing high aridity; and the third one from the assured rainfall zone of Vidarbha, at the foot of hills, with rainfall exceeding 1,400 mm. All the villages face problems of inadequate availability of water for meeting multiple needs such as animal drinking, vegetable cultivation throughout the years and water for basic needs during summer months. Quality and reliability of water are not issues in these villages.

The MUWS vulnerability index was computed for 100 sample rural households in each of the three villages of Maharashtra. Varoshi falls in the high rainfall region, Kerkatta in the low rainfall, drought-prone region and Chikhali in the assured rainfall region. The household-level data on the range of physical and socio-economic variables affecting the MUWS vulnerability were collected through a survey.

In the case of Varoshi, the value of the index was found to be varying from 3.31 to 6.58. Figure 9.1 shows the computed values for all the sample households for all the three villages. Out of the 100 households surveyed, 67 households have vulnerability index values lower than 5.0, which means more vulnerable. In Varoshi, in spite of being a high rainfall region, more households are observed to be vulnerable. This can be attributed to the fact that the main source of water, which is a spring, needs repairs. The villagers were facing water scarcity during 2 months of summer.

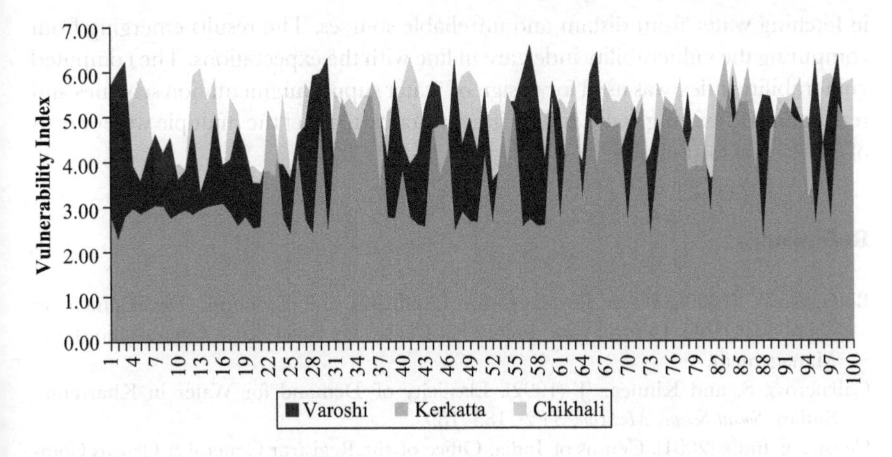

Figure 9.1 Vulnerability Index for Varoshi, Kerkatta and Chikhali Villages

But in the case of Kerkatta village, the computed values of multiple-use vulnerability index at the household level for the sample households range from a low of 2.21 to a high of 6.32 (Figure 9.1). Out of the 100 households, 81 have values lower than 5.0, and therefore are considered vulnerable from a multiple water use point of view. As Kerkatta village falls under low rainfall and drought-prone area in Maharashtra, higher numbers of households are vulnerable.

On the other hand, in Chikhali village, which is an assured rainfall area, the computed values of multiple-use vulnerability index at the household level for the sample households range from a low of 3.15 to a high of 6.37. Out of the 100 households, 30 have a vulnerability index lower than 5.0, and hence are treated as highly vulnerable.

Hence, out of these three villages, Kerkatta had the maximum percentage of households vulnerable to problems associated with inadequate water supply for multiple uses.

Conclusions

The development of an index that helps assess vulnerability of rural population to problems associated with lack of water for domestic and productive water needs is useful in water supply surveillance. In this research, we have developed a household-level MUWS vulnerability index to assess the problems associated with lack of water for domestic and productive needs for rural communities, and subsequently computed its values for 100 sample households, each from three villages in Maharashtra. For this, primary data were collected through a household survey. Among the three villages, Kerkatta in Latur district had the highest number of vulnerable households in the sample. Identification of vulnerable households helped us suggest improvements in water supply to reduce the hardships faced by women in fetching water from distant and unreliable sources. The results emerging from computing the vulnerability index are in line with the expectations. The computed vulnerability index was used for design of water supply augmentation schemes and retrofitting of existing water supply infrastructure to meet the multiple water needs of poor rural households.

References

Botkosal, W. (2009). Water Resources for Livelihood and Economic Development in Cambodia. Solo, Central Java, Indonesia: Centre for River Basin Organizations and Management.

Cairncross, S. and Kinnear, J. (1992). Elasticity of Demand for Water in Khartoum, Sudan. *Social Science Medicine, 34*(2), 183–189.

Census of India (2001). Census of India, Office of the Registrar General & Census Commissioner, New Delhi, India.

DFID (2001). *Addressing the Water Crisis: Healthier and More Productive Lives for Poor People.* London: Department for International Development.

Gelinas, Y., Randall, H., Robidoux, L. and Schmit, J. P. (1996). Well Water Survey in Two Districts of Conakry (Republic of Guinea) and Comparison With the Piped City Water. *Water Research, 30*(9), 2017–2026.

Gleick, P. H. (1997). Human population and water: Meeting basic needs in the 21st century. In R. K. Pachauri and L.F. Qureshy (Eds.), *Population, Environment, and Development.* Tata Energy Research Institute (TERI), New Delhi, India, pp. 105–121.

Howard, G. (Ed.). (2002). *Water Supply Surveillance: A Reference Manual.* Loughborough, United Kingdom: WEDC.

Howard, G. (2005). Effective Water Supply Surveillance in Urban Areas of Developing Countries. *Journal of Water and Health, 3*(1), 31–43.

Howard, G. and Luyima, L. G. (1999). Urban Water Supply Surveillance in Uganda. In J. Pickford (Ed.), *Integrated Development for Water Supply and Sanitation* (pp. 290–293). Proceedings of the 25th WEDC Conference, Addis Ababa, Ethiopia. Loughborough, United Kingdom: WEDC.

Human Development Report (2006). *Human Development Report – 2006.* New York: United Nations.

Hunter, P. R. (2003). Climate Change and Water-Borne and Vector Borne Diseases. *Journal of Applied Microbiology, 94*, 37–46.

Lloyd, B. and Bartram, J. (1991). Surveillance Solutions to Microbiological Problems in Water Quality Control in Developing Countries. *Water Science and Technology, 24*(2), 61–75.

National Sample Survey Organization (1999). Drinking Water, Sanitation and Hygiene in India. NSSO 54th Round, National Sample Survey Organization, Department of Statistics, Government of India.

Nicole, A. (2000). *Adopting a Sustainable Livelihoods Approach to Water Projects: Implications for Policy and Practice.* Working Paper 133. London: Overseas Development Institute.

Rahman, A., Lee, H. K. and Khan, M. A. (1997). Domestic Water Contamination in Rapidly Growing Megacities of Asia: Case of Karachi, Pakistan. *Environmental Monitoring and Assessment, 44*(1–3), 339–360.

Renwick, M. (2008). Multiple Use Water Services. Nairobi, Kenya: Winrock International, GRUBS Planning Workshop.

Sullivan, C. (2002). Calculating Water Poverty Index. *World Development, 30*(7), 1195–1211.

van Koppen, B., Moriarty, P. and Boelee, E. (2006). Multiple Use Water Services to Advance the Millennium Development Goals. IWMI Research Report 98. Colombo: IWMI.

WHO and UNICEF (2000). *Global Water Supply and Sanitation Assessment (2000) Report.* Geneva/New York: World Health Organization/United Nations Children's Fund.

World Health Organization (1976). *Surveillance of Drinking-Water Quality*. Geneva: Author.

World Health Organization (1993). *Guidelines for Drinking-Water Quality: Volume 1 Recommendations* (2nd ed.). Geneva: Author.

World Health Organization (2002). *Reducing Risks, Promoting Healthy Life, World Health Report 2002.* Geneva: Author.

World Health Organization (2004). *Guidelines for Drinking-Water Quality: Volume 1 Recommendations* (3rd ed.). Geneva: Author.

10 The decade of sector reforms of rural water supply in Maharashtra

Nitin Bassi, M. Dinesh Kumar, V. Niranjan and K. Siva Rama Kishan

Background

According to the 2011 census, Maharashtra has nearly 23.8 million households, with 13 million living in rural areas and 10.8 million households in urban areas. About 68% of them have access to tap water sources and 50% to improved latrines. The corresponding figures in rural areas are about 50% and 35%, respectively. Obviously, the progress in the rural areas is far from being satisfactory. But still these figures are a marked improvement from the year 2001, when only about 45.5% and 18% of the rural households had access to tap water and any type of latrine facility, respectively. Further, there is some improvement in the proportion of rural households having drinking water available within their premises (increased from 38.9% in 2001 to 42.9% in 2011).

In spite of the major achievements by the state of Maharashtra in the coverage of rural water supply during the last decade, a large part of the rural population is devoid of access to improved water sources. Further, a significant portion of the water supply infrastructure created performs far below its design potential. This is a reason for slippages experienced by a significant proportion of habitations (World Bank, 2008). As per National Rural Drinking Water Programme database (Ministry of Drinking Water and Sanitation, GOI), around 4,000 rural habitations in Maharashtra slipped back from the water supply coverage status. The major reasons were: 1) drying up of water source, 2) aged nonperforming systems and 3) dwindled supply of water at delivery point. Other reasons identified were poor quality of supplied water, poor operation and maintenance (O&M) and shortage of electricity to operate the systems. Until April 2012, as many as 1,671 habitations, with a total population of 3.5 million people, had their water sources affected by quality deterioration.

The World Bank (2008) report highlights that

> substantial expenditure is incurred by the Government of India (GOI) on rural water supply during the last decade. But, very little is known on how effective this expenditure had been in providing safe water to rural population. Also, there is hardly any analysis of the cost of water supply schemes, cost recovery and subsidies, and the impact of technology choice and institutional arrangements on the level of service. (p. 7)

Pattanayak et al. (2007) also observed that there were only a few rigorous scientific impact evaluations showing that Rural Water Supply and Sanitation (RWSS) policies were effective in delivering many of the desired outcomes. Their analysis showed that RWSS policies are complex with multiple objectives; use inputs from multiple sectors; provide a variety of services (water supply, water quality, sanitation, sewerage and hygiene) using a variety of types of delivery (public intervention, private interventions, public private partnerships, decentralized delivery, expansion or rehabilitation); and generate effects in multiple sectors (water, environment, health, labor). The fact that the communities are mostly dependent on multiple sources of water supply, including informal sources, makes evaluation of policy impacts often complicated.

In order to understand the impact of reforms, including changes in legal, institutional and policy framework, on drinking water sector performance, it is important to examine whether there has been a major change in technology choices for water supply and any enhancement of institutional capacities available for managing the schemes at various levels, along with changes in performance vis-à-vis level of service – coverage, water supply levels, water quality, frequency and reliability – cost per unit volume of water supplied and cost recovery. Thus, this chapter analyzes the institutional and policy framework for rural water supply in Maharashtra, with particular reference to the 'Sector Reform' initiated by the government in 1999, to see how they influence the performance of rural water supply schemes.

The era of reforms

Till the mid of ninth 5-year plan (1999), approach to the provision of rural water supply in India was supply driven, with emphasis on norms and targets and on construction and creation of assets, and with relatively little concern for suitable arrangements for the less attractive but critical issue of better management and maintenance of the facilities already built (WSP, 2004). The water supply requirements of the scheme were decided on the basis of a few simplistic norms on the per capita water requirements of the rural households and the population to be covered. Consequently, no attention was paid to the socio-economic characteristics of the population to be served, the climatic conditions or cultural aspects, which would have a major bearing on the actual water needs in rural areas for domestic purposes. Further, the productive water needs were completely ignored.

In the Accelerated Rural Water Supply Programme (ARWSP), effort was only on increasing coverage, with little concern for the sustainability of the technical system created and the resource base that these systems tapped. Because of this poor vision of water resources management and O&M of the systems, at several places rural water supply infrastructure rarely lasted its designed life and the service quality remained far below the expectation.[1] Inefficient service delivery, poor quality of infrastructure and poor maintenance of the system under the supply-driven approach led to a fundamental policy shift in the sector by the end of the 1990s, toward an approach that was widely acclaimed as 'demand-driven approach' to rural water supply. Under the new approach, community participation and decentralization of powers for implementing and operating drinking water supply was made an important prerequisite. It was envisaged that the role of government should be that of a facilitator. This was to be achieved through the Sector Reforms Project (SRP), which was launched in 1999 on a pilot basis in 67 districts across the country.

By 2003, the SRP was scaled up to cover the entire country under the Swajaldhara Programme. It envisages 10% of the capital cost and 100% of the O&M cost to be borne by beneficiaries, with the central government providing the rest. However, the limited external assessments with regard to the implementation of the Swajaldhara scheme showed that there were serious pitfalls in the proposed reform principles (Sampat, 2007). The government agencies responsible for implementation of drinking water supply policy were not keen on bringing about the changes that were required to hand over some of the powers, like planning and operation of drinking water schemes, to local communities. Further, the principle of seeking community contribution toward capital cost was politically unappealing for the state governments (Cullet, 2011). As a result, in many villages, Swajaldhara projects did not follow the envisaged guidelines. During the 11th 5-year plan, a new policy instrument, i.e., National Rural Drinking Water Programme (NRDWP), was formulated. The NRDWP emphasized the need to conceive drinking water supply in a wider context of public health. Thus, it provides a direct link with sanitation. Further, it proposed links with health policy and also with the Mahatma Gandhi National Rural Employment Guarantee Scheme (MGNREGS) (Cullet, 2011). However, attempts to integrate with other programs such as MGNREGS had undesirable consequences. MGNREGS is strictly an employment guarantee scheme, which aims at providing 100 days of unskilled manual work to every rural household. Its role in building water supply schemes, which require both skilled labor and engineering infrastructure, is too insignificant. In many instances, this led to local government agencies undertaking desilting of ponds, tanks and check dams in the name of 'water supply improvement or augmentation,' only to tap MGNREGS funds.

In Maharashtra, a demand-responsive approach was implemented for delivery of rural water supply and sanitation services in 2000. Prior to that, in the early 1980s, there was the emphasis on individual village-based schemes, which mostly tapped local groundwater sources. Regional rural water supply schemes increased in the late 1980s, and particularly, during the mid to late 1990s. These regional supply schemes drew water from large reservoirs, and were considered to be better than the individual water supply schemes based on underground water, as they involved economies of scale and the water was of better quality (Sangameswaran, 2010). Between the mid-1980s and 1990s, the state government implemented various schemes for improving the water supply coverage with the financial support from the World Bank, the German Development Bank and DFID. In 1995, Maharashtra was the first state in the country to prepare a White Paper on the water situation in the state and to initiate institutional reforms addressing improvement of the performance of local bodies that are responsible for provision of drinking water and sanitation facilities (Das, 2006). The White Paper primarily addressed the issues of disparity in water supply, level of service and groundwater depletion. A few recommendations were made concerning water supply in rural areas and enforcement of various legislations, stressing upon more effective management of water supplies and decentralization.

For drinking water, a master plan consisting of a number of detailed proposals was adopted in 1996. The overall goal of the proposals was to make Maharashtra free of tankers – a major source of drinking water for villagers during summer months – by the year 2000. As per the master plan, water supply norm became more rigid, with supply rate increasing from 40 to 55 lpcd. More importantly, a source dependability criterion of 95% was adopted for water sources. This meant that the surface water schemes had to be taken up in a big way, as groundwater-based schemes were found to be less dependable, with sources drying during summer months. Thus, the regional supply schemes acquired prominence. This led to the establishment of a parastatal agency called the Maharashtra Jeevan Pradhikaran (MJP) (Sangameswaran, 2010). The MJP built, tested and then handed over schemes to the concerned local authorities to operate and maintain. For each completed scheme, MJP charged a commission of 17.5% on the total costs to meet its overhead costs and other administrative expenses. However, the tanker-free program was not a complete success. It required huge financial outlay, and the regional supply schemes brought in their own problems, which include high capital costs, lack of willingness of local authorities to take over the schemes and inequity in distribution of water between head-end and tail-end villages (Sangameswaran, 2010).

With the change of the state government in 1999, the centrally sponsored SRP was initiated in four districts of the state. By the year 2000, in principle the role of government was shifted from service provider to that of policy formulation and capacity builder. During the reform process, communities were encouraged to decide the water supply schemes of their choice, and efforts were made to ensure their sustained involvement through cost sharing. Decisions on the type of scheme; the implementing body, i.e., whether it should be the gram panchayat (GP), the Zilla Parishad (ZP) or a nongovernmental organization (NGO), as well as the technical service provider (whether it should be MJP, the ZP or a private party) rested with the Grama Sabha (GS) at least nominally. These ambitious reforms were in a sharp contrast to the pre-2000 situation, where both the need for drinking water and the manner of its provision were determined by state departments and parastatal agencies such as MJP, ZPs and Panchayat Samitis.

Further, a major role is to be played by village water and sanitation committees (VWSCs), which are elected by the GS and are technically subcommittees of the GP. They would have funds directly devolved to them, and maintain accounts separately (from the general accounts of the GP) and plan and implement the scheme autonomously (Sangameswaran, 2010). Further, during this phase there was a technological shift from regional supply schemes based on reservoirs to individual-village schemes based on local groundwater sources. It was believed that mobilizing local community and costs recovery, which are important components of the supposedly 'demand-driven approach,' would be easier in small individual village-based schemes.

In 2002, German KFW funded a 'demand-driven' scheme, called *Aaple Pani*, which was initiated in three districts of the state. In the year 2003, the World Bank financed Maharashtra Rural Water Supply and Sanitation Project, also called *Jalswarajya*, which was initiated to cover rural areas of remaining 26 districts of the state. The overall objectives of Jalswarajya were to increase access to RWSS services and to institutionalize decentralized delivery of RWSS services by local governments (RSPMU, 2004). Jalswarajya relied on voluntary participation by communities, wherein communities selected water supply and sanitation services from a menu of options, and targeted provision of RWSS services by project administrators (Pattanayak et al., 2007). As per the project guidelines, community contribution to the capital cost was kept at 5% for tribal villages and 10% for nontribal villages; and communities were also required to bear 100% of the O&M costs. The rest of the capital cost was met by the project.

World Bank (2011) observed:

> Jalswarajya is recognized both in the area of providing access to water and also for its initiatives in the sanitation sub-sector. Most of the houses in the project area now have water connection and a toilet. Further, more than 1800 GPs in the project area have been declared free of open defecation. (p. 4)

However, because of these externally financed schemes, central government–assisted Swajaldhara and NRDWP made modest progress in the state. Until May 2012, only Rs 12.79 crore was spent by the state against the total central release of Rs 101.31 crore under the NRDWP (Ministry of Drinking Water and Sanitation, GOI). It seems that the greater focus on community mobilization and substantial financial allocation for capital creation made these externally financed schemes more popular with both rural communities and the state authorities.

Institutional framework

Legal and policy framework for RWSS

GOM established the Ground Water Survey and Development Agency (GSDA) to scientifically tap the groundwater resources in 1972. Soon after, the Maharashtra Water Supply and Sewerage Board Act (MWSSB Act) was passed, in 1976. Under the Act, Maharashtra Water Supply and Sewerage Board (MWSSB) was set up in 1977 to take over the functions and assets of the Public Health Engineering (PHE) department of GOM.

Due to groundwater overexploitation and resulting water scarcity in many areas during the 1990s, a large number of settlements in Maharashtra had to depend on tankers to meet their drinking water requirements and more so during the summer months. In order to manage the exploitation of groundwater for the protection of public drinking water sources, the Maharashtra Groundwater (Regulation for Drinking Water Purposes) Act, was passed in 1993. The Act prohibited the construction of wells within a radius of 500 m from a public drinking water source, if both are in the area of the same watershed. Further, the Act empowered the appropriate authority (district collector in this case) to restrict or prohibit extraction of water (for any purpose other than for drinking) from any well in an identified 'water-scarce' area (as advised by GSDA) during the scarcity period, if such well is within a distance of one kilometer of the public drinking water source (GOM, 1993). However, as noted by Phansalkar and Kher (2006), the Act was not preventive in nature, but only corrective. For instance, the provisions of the Act were only enforceable either in watersheds declared as 'overexploited' or if a specific locality was notified as scarcity affected in a particular year. There were no provisions for registration of wells or for making applications mandatory for sinking new wells. It did not even provide for compulsory licensing of drilling companies or agencies.

The White Paper on the state's water situation attributed the problem of drinking water scarcity to inadequate infrastructure development and to the excessive dependence on unreliable sources. Further, it identified the need for massive capital investments for developing the required infrastructure for fulfilling the drinking water requirements. As a result, the state government amended the MWSSB Act and established the Maharashtra *Jeevan Pradhikaran* (MJP), a statutory body constituted from erstwhile MWSSB in 1997, giving the state the authority to raise capital from the open market. With the help of the MJP and the GSDA, the government of Maharashtra embarked on a mission to provide sustainable water supplies for both urban and rural areas. In 2003, Maharashtra became one of the few states in India to adopt a state water policy. It laid emphasis on management, operation and maintenance of these services by community-level organizations and appropriate local-level bodies (GOM, 2003).

Of late, GOM has passed the Maharashtra Groundwater (Development and Management) Act of 2009, which replaced the earlier Groundwater Act of 1993. This new Act is more holistic and aims at

> facilitating and ensuring sustainable equitable and adequate supply of groundwater of prescribed quality, for various category of users, through supply and demand management measures, protecting public drinking water sources and establishing the State Groundwater and District Level Authority to manage and to regulate, with community participation, the exploitation of ground water within the State of Maharashtra. (GOM, 2009, p. 1)

As per section 3(1) of the Act, the Maharashtra Water Resources Regulatory Authority (established under section 3 of the Maharashtra Water Resources Regulatory Authority Act, 2005) shall be the State Groundwater Authority. GSDA has also been provided with more footholds under the Act.

In contrast to the Groundwater Act of 1993, which empowered the district collector (in consultation with a technical officer) to notify the area as 'overexploited' or 'water scarce' (and that too in an ad hoc manner), the new Groundwater Act empowers the state groundwater authority to notify an area but only on the basis of recommendations from the GSDA, views of various institutions working in the groundwater field and views of the users of the groundwater of the area. The decision to notify an area has to be based on scientific studies on groundwater balance and quality and groundwater estimation. Further, the Act calls for establishment of a Watershed Water Resources Committee (WWRC) to promote and regulate the development and management of groundwater in the notified area. For the notified areas, the Act envisages several restrictions such as bans on the construction of wells, prohibition on groundwater pumping from the existing deep wells (more than 60 meters deep) and stipulation on deep well users to follow the groundwater use plan and crop plan. All these measures are now in operation in notified areas. Unlike the earlier Act, the new Act puts emphasis on the protection and preservation of groundwater quality of all the existing drinking water sources in the state.

Further, in both notified and unnotified areas, registration of all the well owners is made mandatory (section 7 of the Act); and drilling of deep wells for agriculture and industrial use is prohibited (section 8.1 of the Act). Additionally, section 12 of the Act made registration compulsory for drilling rig owners and operators in the state. The Act also empowered the state authority to identify the potential areas for recharge schemes, in consultation with the GSDA and the Central Ground Water Board (GWB). The Groundwater Act of 2009 also empowers the District Authority (officer not below the rank of *Tahsildar*) to enforce the decisions of WWRC. Whenever necessary, the District Authority or any officer duly authorized by it, after giving prior notice to the owner or occupier of any land, may initiate an inquiry or implement or enforce any decisions in connection with the protection of a public drinking water source or with the maintenance of a public water supply system. The District Authority can seize any equipment or device utilized for illegal sinking or construction and can demolish the executed work either fully or partly. Further, the District Authority can also direct any groundwater user who does not comply with the provisions of this Act and rules framed thereunder to close down the extraction of groundwater, and can temporary disconnect its power supply and seal any hydraulic work that is found to be illegal. Although the Maharashtra Groundwater (Management and Development) Act of 2009 is a major improvement over the earlier Act, its effectiveness in arresting groundwater exploitation can only be judged once it is implemented across the state.

Government resolutions and supply norms

Between 2000 and 2004, Water Supply and Sanitation Department (WSSD) passed seven Government Resolutions (GRs) to operationalize the RWSS reform policy across the State. These GRs dealt with: 1) establishing 40 lpcd as the water supply norm and focusing on community participation in Rural Water Supply Programme (RWSP), 2) implementation of sanitation reforms, 3) panchayati raj institutions

(PRIs) and their role in RWSP, 4) augmenting water supply and strengthening of drinking water sources, 5) formation of social audit committees, 6) having competent authorities for source certification for RWSS schemes and 7) availing services of private service providers for finalization of water sources.

From 2008 onward, around 20 different GRs mostly relating to the administrative, financial and managerial aspects of rural water supply have been passed by the GOM.

Cost norms

The GOM prescribed cost norms for implementing piped water supply schemes across the state (Table 10.1). These cost norms were recommended in 1999 and have not been revised since then. No provision for inflation was kept that could be consulted while planning the new schemes (Gawade, 2012).

These cost norms are for the new piped water supply schemes, but they are also used for augmentation of the existing scheme, if the cost is lower than the new scheme. In the case of schemes where the per capita cost is above the standard norms, separate approval at the state level is required. Field experiences suggest that the present cost norms are insufficient to cover the current capital cost of the schemes and need to be revised (Gawade, 2012). To avoid approvals at the state level and delays in implementation, the current practice is to limit the scheme components in a way that the total cost remains below or at par with the prescribed standard norms. This approach has two implications: First, the new or augmented schemes may not be able to cover the entire population of the habitation under consideration; and second, sustainability (i.e., drinking water security at the household level) may not be ensured (Gawade, 2012). It is important to mention here that NRDWP has kept around 20% of the total allocation toward its sustainability component but does not suggest any cost norm. Further, 100% of these funds will be provided by the central government. Thus, it is high time now that the state should revise the present cost norms to better serve the uncovered or partially covered, quality affected and slipped-back habitations.

Table 10.1 Cost Norms for Implementing Piped Water Supply Schemes in Maharashtra

Type of Scheme	*Cost Norms in INR per Capita*	
	Non-Konkan Area	*Konkan Area*
Hilly areas	2,120	2,320
PWS with static lift of more than 30 m	1,790	1,970
PWS with static lift up to 30 m	1,390	1,530

Source: World Bank (2012).

Organizations

The Water Supply and Sanitation Department (WSSD), GOM, is the state nodal agency for formulating, implementing, operating and maintaining regional water supply schemes in both rural and urban areas. The GSDA and the MJP are the two technical line agencies supporting the WSSD.

The GSDA is a technical agency (mostly geologists), and is entrusted with the responsibility of overall development and management of groundwater. Its directorate is located in Pune, which is assisted by six regional and 33 district-level offices. Senior geologists at district level (who are supported by junior engineers) report to deputy directors at the regional level, who are ultimately responsible to GSDA directorate. Details of the staff categories are provided in Table 10.2.

The MJP mainly consists of engineers, and implements the piped water supply schemes. The MJP, with central offices in Mumbai and Navi Mumbai, has field offices spread across the entire state. Overall, there are five zonal offices, 16 circle offices, 44 work/project divisions and 151 subdivisions. The Member Secretary, who is the Chief Executive Officer of the MJP, has the status of Secretary to Government. The Member Secretary (Technical) has technical control over field works, and is the person to whom the Superintending Engineers for coordination, central planning, designing and monitoring report. There are also other subordinate officers and staff. Overall staff strength of MJP is 7,186, with 1,575 (21.9%) technical, 2,947 (41%) nontechnical and 2,664 (37.1%) contract-based positions (Figure 10.1). The primary responsibility of MJP is planning, design, investigation, detailed engineering and execution of water supply and sewerage schemes in the state. Additionally, MJP arrange finances for these schemes. On the successful completion of these projects, MJP hands them over to the respective local

Table 10.2 Sanctioned Post vs. Position Filled Under Various Staff Categories, GSDA

Staff Category	*GSDA*	*Zilla Panchayat (ZP)*
A	Senior management, senior geologists, senior chemist, senior administration officials	Deputy engineer, assistant and junior geologists
B	Deputy engineer, assistant and junior geologists, assistant chemist, analyst, planning officer, assistant administrative	Junior engineer, rigman, driller, compressor operator, clerks, drivers
C	Junior geophysicist, junior chemist, assistants and clerks	Assistants and peons
D	Peons, helpers, watchmen	Peons, helpers

Source: World Bank, 2012

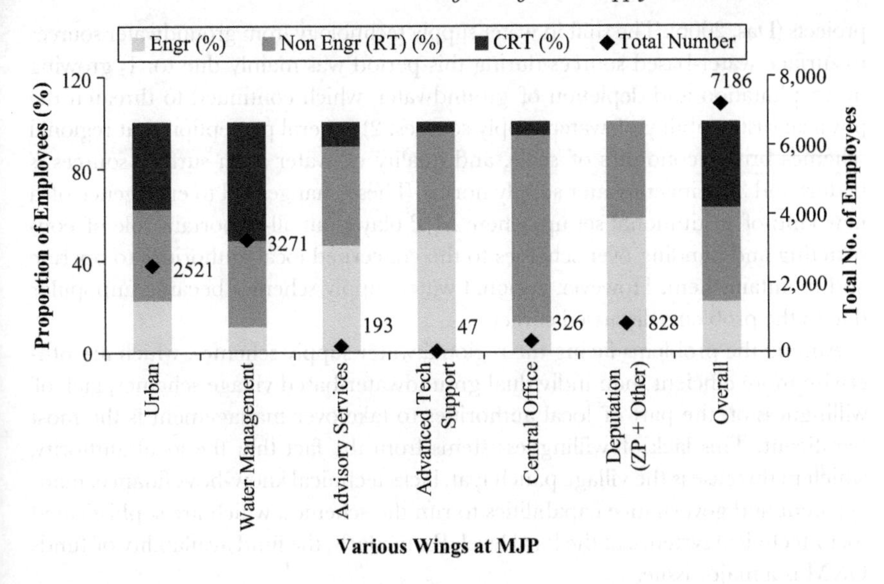

Figure 10.1 Employee Proportion in the Various Wings at MJP

Source: Information provided by MJP office, Mumbai

bodies. To settle the administrative expenses, MJP receives a fixed amount on total project costs which has been currently fixed by GOM at 17.5% of the value of projects.

A Reform Support and Project Management Unit (RSMU) was also set up in the rural drinking water department of ZP to facilitate the RWSS reforms process. Additionally, GP is made responsible for demanding new drinking water supply schemes from the ZP. GS is also empowered to decide on the kind of scheme, the implementing body, as well as about the technical service provider. In each village, VWSC, a subcommittee of the GP, was formed to plan, implement, operate and maintain the village water supply scheme autonomously.

Water and Sanitation Support Organization (WSSO) was established to function as Directorate of Water Supply and Sanitation under the WSSD and was made accountable to the State Water and Sanitation Mission. GSDA and MJP were entrusted with the responsibility of providing technical support to WSSO.

Water supply technologies and institutional capacity

Until 1985, the majority of the rural water supply schemes in Maharashtra were based on either dug wells or bore wells. The period from 1985 to 1997 witnessed an increase in the number of piped water supply schemes, which were based on surface water sources. During 1991 to 1998, 50 regional schemes, mostly based on surface water sources, were implemented under the various externally supported projects (Das, 2006). The shift in water supply technology from groundwater sources to surface water-based sources during this period was mainly due to: 1) growing overexploitation and depletion of groundwater, which continued to threaten the physical sustainability of water supply sources; 2) general perception that regional schemes bring economies of scale, and quality of water from surface sources is better; and 3) stringent water supply norms. These changes led to emergence of a new kind of institutional set up, where MJP played an all-important role of constructing and handing over schemes to the concerned local authorities to operate and maintain them. However, regional water supply schemes became unpopular due to the problems discussed earlier.

Among the problems facing the regional water supply schemes, which are otherwise more efficient than individual groundwater-based village schemes, lack of willingness on the part of local authorities to take over management is the most significant. This lack of willingness stems from the fact that the local authority, which in this case is the village panchayat, lacks technical know-how, finance, management and governance capabilities to run the schemes, which are sophisticated socio-technical systems, at the local level. Particularly, the fund availability of funds O&M is a major issue.

A recent evaluation of WATSAN sector in India pointed out that decentralization is also fraught with problems at the local level due to issues of 'accountability.' In the case of Maharashtra, MJP takes up execution of rural water supply schemes on requests from concerned Zilla Parishads/GPs and/or when the rural scheme cost is equivalent or above Rs 50 lac. On completion of such schemes, MJP is expected to hand over the scheme to ZP (in case of multi-village schemes) and GP (in the case of small schemes) for operation and maintenance. In such cases, it receives the funds directly from the WSSD (Gawade, 2012). Hence, it has no special incentive to make sure that the system so built runs reliably and efficiently.

As a result, individual village-level schemes (mostly based on groundwater sources) of water provision were again looked upon as better alternatives. Along with this change in the scale of organization of water systems came the shift from a supply-driven to a demand-driven approach (Sangameswaran, 2010). Under this new approach, the Gram Sabha and VWSC were given the authority to decide on the kind of scheme, and also to implement, operate and maintain it. The official authority of the state government departments (including MJP and ZP) was minimized. However, this demand-driven decentralization approach remained only on paper.

This is evident from the fact that out of 13,515 completed schemes, only 1,951 schemes were handed over to GP. Most of the success in terms of community contribution to the capital cost of the water supply infrastructure came only in the schemes that were externally aided and had a huge financial allocation for the community mobilization. Further, the inability of the VWSC to encourage users to pay a water fee affected the overall financial recovery, which also adversely impacted on the overall performance of the schemes. Until March 2007, the dues outstanding in respect of execution and maintenance of rural piped water supply schemes from ZP/GP amounted to about Rs 504 crore. Further, the dues outstanding with respect to water charges are major (72%) and those related to community (popular) contribution are minor (only about 4%). Moreover, as almost 80% of the implemented schemes during the reform era tapped groundwater, the ability of such sources to supply water in summer or during drought years was doubtful.

Performance of selected water supply schemes

A total of 12 rural water supply schemes representing six different regions of Maharashtra were selected to analyze the scheme performance. The sampling of schemes and villages ensured different types homogeneity of geo-hydrology and other scientific parameters.

The management and institutional performance of the utilities were studied by analyzing data collected from the managers of the schemes and communities that are served by them on various physical, socio-economic and institutional factors, which determine these performance variables. The water administration was studied by analyzing the data from the state-level and local-level bureaucracies concerned with drinking water supply.

Water supply coverage

Among selected schemes, 45% of the rural households were found to be covered by formal water supply in the selected schemes. There are significant differences in coverage across water supply technologies. The highest coverage was achieved in a single-village scheme based on bore well (94.5%), followed by multi-village schemes based on surface reservoir (80.5%), and a single-village scheme based on river lifting (74.2%) and reservoir (73.7%). The regional scheme based on river lifting had the lowest coverage (Table 10.3). As Table 10.3 shows, the single-village schemes based on dug wells, percolation wells and infiltration wells also have relatively lower percentage coverage as compared to reservoir-based single and multi-village schemes (42.7%, 47.3% and 60.4%, respectively). Notably, there is a remarkable difference in coverage between schemes that tap surface reservoirs and that lift river water directly, when the scheme has to supply water for a large number of villages (73.8% against 21.9%). Differences in coverage were also seen with change in type of source. In the case of single-village schemes, those tapping water from surface sources (73.8%) are found to be performing better than those tapping groundwater (64.2%) and subsurface water (57%).

The number of individual tap connections was highest in the single-village scheme based on river lifting and surface reservoir (74.3% and 73.7%, respectively) and lowest (35.6%) for the single-village scheme based on subsurface water. Nearly 63% of rural households that are covered by regional water supply schemes based on reservoirs also had individual household tap connections. With regards to type of source, single-village schemes tapping water from surface sources are found to have the highest percentage of individual tap connections (73.7%), followed by those tapping groundwater sources (48.2%), and those tapping subsurface water had the lowest percentage of households with individual tap connection (35.6%).

Table 10.3 Households Covered With Water Supply Schemes Under Different Techno-Institutional Arrangements

Type of Scheme and Source	*Technology*	*Districts From Which Schemes Are Selected*	*Average Number of Households/ Scheme*	*Average Number of Households Covered/ Scheme*	*Overall Percent of Households Covered*	*Overall Percent of Households Covered With Individual Piped Connection*
Individual GP (Ground water-based)	Bore well	Nandurbar	555	522	94.1	55.9
	Dug well	Bhandara and Amravati	384.5	164	42.7	42.7
	Overall	–	**441.33**	**283.3**	**64.2**	**48.2**
Individual GP (Subsurface)	Infiltration well	Hingoli	508	307	60.4	60.4
	Percolation well	Thane	1,059	501	47.3	23.6
	Overall	–	**783.5**	**404**	**51.6**	**35.6**
Individual GP (Surface water–based)	Surface reservoir	Jalgaon	1,140	840	73.7	73.7
	River lifting	Sindhudurg	132	98	74.2	74.2
	Overall	–	**636**	**469**	**73.8**	**73.7**
Regional (Surface water–based)	Surface reservoir	Gondia, Yavatmal, Osmanabad and Pune	9,904.67	7,223	80.50	62.9
	River lifting	Solapur	50,275	11,005	21.9	–
	Overall	–	**19,997.25**	**7,979.4**	**43.7**	–

Source: Authors' own analysis using primary data collected from various water supply agencies

Overall, it appears that the water supply schemes based on surface reservoirs, both single-village schemes and multi-village schemes, show relatively better performance as compared to single-village schemes tapping groundwater, and subsurface water, and regional schemes lifting water directly from the river, in terms of the proportion of the households actually served by them and proportion of households provided with individual tap connections.

Adequacy of water supplies

All the schemes were found to be supplying at par or above the state-adopted norm of 40 lpcd per capita in all seasons, except in the case of one individual scheme (tap subsurface flow using infiltration well) that supplied only 32 lpcd during summer months. Overall, the average per capita water supply is slightly higher for groundwater-based schemes than their counterparts based on reservoirs and river lifting. But this is because in open well-based schemes, an excessively high level of supply is achieved (110 to 117 lpcd) primarily due to a small percentage of the targeted households (42.7%) being served.

In the case of single-village schemes based on groundwater, percolation wells and infiltration wells, the average level of water supply per capita is slightly better due to the extent of coverage of the targeted households. For instance, the schemes that tap surface reservoirs supply water to 73.7% of the total households considered in the design, whereas the schemes that extract groundwater and subsurface water supply water to a significantly lower proportion of households (57%).

Within the scheme, no major variation is seen in the frequency of water supply across different seasons. Amongst schemes, frequency of water supply ranged from twice a day to once in 2 days for single-village schemes based on groundwater; twice a day for single-village schemes based on surface reservoir; once a day for individual-village schemes based on river lift; to once a day to once in 2 days for regional schemes based on both surface reservoir and river lift. In terms of hours of daily water supply, regional schemes based on surface water fare much better than any other schemes (Figure 10.2). Such schemes were found to supply water on an average of 7.9 hours during monsoon, 5.6 hours during winter and 6.1 hours during summer. Individual schemes that abstract groundwater using bore wells showed low performance, supplying water only for 1 hour during monsoon and winter; and for 0.67 hours per day during summers. As discussed earlier, sustainability of schemes supplying on groundwater during summer months is uncertain.

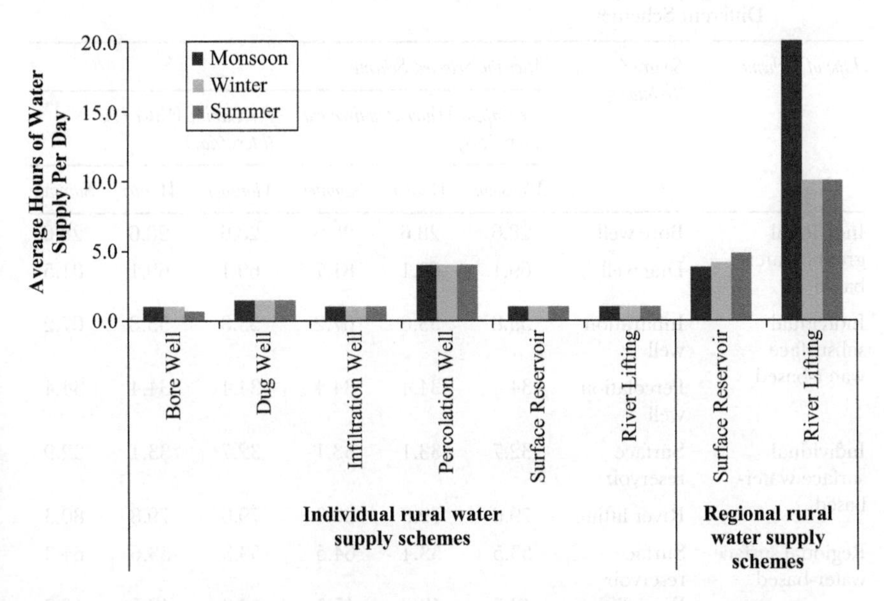

Figure 10.2 Scheme and Source-Wise Average Rate of Water Supply (hours/day)

Source: Authors' own analysis using primary data collected from various water supply agencies

Further, as demand for household water could vary from place to place depending on the socio-economic dynamics of the household and the cultural factors over and above the climatic factors, a final assessment of the performance can be made only if we see what proportion of the actual water demand is being met by the scheme. The analysis of household data shows that the per capita water use is highest for households under single village river lifting schemes (around 80 lpcd), followed by 75 lpcd for dug well-based schemes and 59 lpcd for regional schemes based on reservoirs (Table 10.4). The lowest per capita water use was found in the case of bore well-based schemes (28.6 lpcd). What is interesting is that the per capita requirement was also proportionally low for bore well-based schemes, and that the entire requirement is met by the schemes. One possibility is that due to water shortage, the communities are reducing their water demands by not going for economic activities such as livestock keeping. In other cases, where the water use is high, the communities might have demanded a new scheme to meet their domestic water requirements, like in the cases of the dug well-based scheme, or the single village reservoir scheme or the multi-village scheme.

In the extent of fulfillment of the household water requirement, there is significant improvement with the introduction of new schemes in all except the case of multi-village schemes based on reservoirs. The largest change is found in the case of individual schemes tapping subsurface water. Marginal improvement is also observed for the bore well-based schemes, surface reservoir-based single-village schemes and river lifting-based multi-village schemes. However,

Table 10.4 Average per Capita Water Use Against the Requirements of Households Under Different Schemes

Type of Scheme	*Source/ Technology*	*After the Selected Scheme*					
		Per Capita Water Requirement (liters/day)			*Per Capita Water Use (liters/day)*		
		Monsoon	*Winter*	*Summer*	*Monsoon*	*Winter*	*Summer*
Individual groundwater-based	Bore well	28.6	28.6	28.6	28.6	28.6	28.6
	Dug well	69.1	69.1	81.7	69.1	69.1	81.5
Individual subsurface water-based	Infiltration well	53.8	53.8	67.2	53.8	53.8	67.2
	Percolation well	34.4	34.4	34.4	34.4	34.4	34.4
Individual surface water-based	Surface reservoir	32.7	33.1	33.1	32.7	33.1	32.9
	River lifting	79.6	79.8	80.3	79.6	79.8	80.3
Regional surface water-based	Surface reservoir	53.5	53.4	64.5	53.3	53.6	64.3
	River lifting	43.7	43.9	47.5	43.3	43.5	46.2

Source: Authors' own analysis using primary data collected from the field survey

Table 10.5 Extent to Which Water Requirements Are Met by the Primary Source Before and After Introduction of the Current Water Supply Scheme

Type of Scheme	*Resource Tapped*	*Technology Used*	*Percent of Household's Total Water Requirement Met by Primary Source*		*Percent Change*
			Before Scheme	*After Scheme*	
Individual	Groundwater	Bore well	99.7	100.0	0.3
		Dug well	85.8	85.6	–0.2
		Overall	90.4	90.4	0.0
	Subsurface water	Infiltration well	85.0	95.5	10.4
		Percolation well	83.3	99.0	18.8
		Overall	84.2	97.3	13.1
	Surface water	Surface reservoir	96.7	100.0	3.4
		River lifting	100.0	100.0	0.0
		Overall	98.4	100.0	1.7
Regional Scheme	Surface water	Surface reservoir	87.8	77.6	–11.6
		River lifting	95.6	98.0	2.5
		Overall	89.1	85.8	–3.3

Source: Authors' own analysis using primary data

in the case of the multi-village scheme based on reservoir, the situation deteriorated (Table 10.5). This is partly explained by the fact that nearly 45% of the households depend on a secondary as well as a primary source to meet their household water needs. Nevertheless, in the case of single-village scheme based on surface water, currently 100% of the requirement is met from the new source, as compared to 97.3% for subsurface schemes and 90.4% for schemes based on groundwater.

Quality of water supply

The degree of access to the source and the percentage of households that exclusively depend on the primary source can be good indicators of the quality of water supply from a source. The first one would decide whether the households are able to access water from a particular source with ease or not. Here, we would consider only those who access water from the source within the premise as well as those who have individual household tap connections. The second one would indicate how much importance the source has in meeting the water needs of the households covered by the scheme.

As per the analysis of the collected field data, the proportion of households that depend exclusively on the primary source is higher for surface water-based schemes. Across technologies, in the case of bore well schemes and single-village schemes based on river lifting, all the households depend exclusively on the primary source. On the other hand, none of the households covered by the scheme supplying water through percolation wells depend exclusively on the same. A significantly high proportion of selected households under the surface reservoir-based single-village scheme (98%) depend exclusively on the primary source. In the case of multi-village reservoir-based schemes, nearly 50% of the households depend exclusively on the primary source, and the remaining 45.7% depend on both the primary source and common sources. The importance of the source is also highest for single-village schemes based on surface water, with 98% of the households exclusively depending on the primary source.

On the whole, amongst different water sources, the degree of access to water supply source appears to be much higher for single-village schemes based on surface water (85.2%) as compared to those based on groundwater (62.3%) and subsurface sources (35.3%). The figures, however, change with changes in technology. In the case of bore well–based schemes, the degree of access is the highest (92.3%), followed by single-village schemes based on surface reservoir (86.3%) and river lifting (84%). It is 70.6% for percolation well-based schemes and 54.3% for multi-village schemes based on surface reservoirs. But it is also noteworthy that in the case of bore well–based schemes, the average amount of water used by the households is very less.

An alternative indicator for and perhaps a more accurate way of measuring the performance of a new water supply scheme is the average reduction in distance to the source. As a result of introduction of an improved source of water, the distance traveled by the members of the household to collect water has reduced considerably after introduction of new schemes. Maximum reduction in distance traveled per sample households was found in the surface reservoir-based multi-village scheme, which is followed by surface reservoir-based single-village scheme, subsurface water-based single-village scheme and bore well-based single-village scheme. The least reduction was in the case of regional river lifting schemes.

Capital and operational cost of water supply

Capital cost, in terms of rupees per unit volume of water supplied per scheme, was found to be highest (Rs 0.57 per liter) for regional schemes that are based on surface reservoirs. Lowest was for individual schemes that are based on surface water (Rs 0.10 per liter). Among the schemes that are based on groundwater and subsurface water, the scheme based on percolation wells has the highest capital cost (Rs 0.37) per liter of water supplied, which is followed by the one based on infiltration and bore wells (Rs 0.15/liter each); and dug wells (Rs 0.14/liter). But one needs to use these figures with caution. There are many reasons for this. First, the life of the system would differ drastically with changing type of scheme. Although the water distribution infrastructure would have the same life across schemes, the life of the source would change across resource types. A large or medium reservoir that taps surface water from a large catchment would generally have a long life. At the same time, a well, which taps aquifer in the hard rock area, would have a very short life of 10–12 years. Second, wells are not substitutes for regional water supply schemes. Instead, RWSSs are resorted to when local groundwater-based sources fail due to droughts or resource depletion. Since they address the specific problem of source sustainability, comparison of costs between such schemes would be inappropriate.

As regards the average operational cost per household for the households covered by formal water supply, the highest was for the regional schemes based on river lifting (Rs 5,467), followed by single-village schemes based on river lifting (Rs 1,190). One reason for the exceptionally high O&M cost for the regional water supply scheme based on river lifting is that a relatively small proportion of the households considered for design of the scheme actually benefit from it. This could be due to poor yield from the catchment. The average operational cost per household was lowest for the bore well-based scheme and schemes based on infiltration wells (Rs 322) and percolation wells (Rs 369). Interestingly, for the reservoir-based multi-village scheme, the average O&M cost per year was only Rs 1,138 per household, which is slightly lower than the O&M cost for the dug well-based scheme (Rs 1,181). This is contrary to the general perception that regional water supply schemes incur high operation and maintenance costs. The O&M cost was much lower for the single-village scheme based on surface reservoir (Rs 445), which is less than half that of the dug well-based scheme.

But to understand what factors contribute to the cost differences, it is necessary to look at the break up of these cost figures. We can see that the average maintenance and repair (M&R) cost per household (covered by the scheme) was the lowest for percolation well-based scheme (Rs 50/household/annum), followed by infiltration well-based scheme (Rs 65/household/annum). M&R was significantly higher for the individual-village scheme based on wells (Rs 284/household/annum). For surface reservoir-based schemes, the values were Rs 86/household/annum to Rs 140/household/annum, for single-village schemes and regional water supply schemes, respectively.

Average salaries per covered household were also quite high in groundwater-based individual schemes (Rs 373.4), next only to surface water-based regional water supply schemes (Rs 473.3). Further, average electricity charges were significantly higher for regional schemes based on river lifting (Rs 1154.6/household covered), as water transport and distribution involved many stages of pumping to take water to high elevations. The electricity charges as a percentage of the annual O&M cost varied from 15 for single village bore well-based schemes to the highest of 67 for regional water supply schemes based on river lifting. We can see that the electricity charges significantly contributed to the differences in O&M cost across scheme types.

Recovery of water tariff

Average water tariff per household connection varied from Rs 30/month/connection in bore well-based groundwater schemes to Rs 75/month/connection in surface reservoir-based regional schemes. For other covered households under

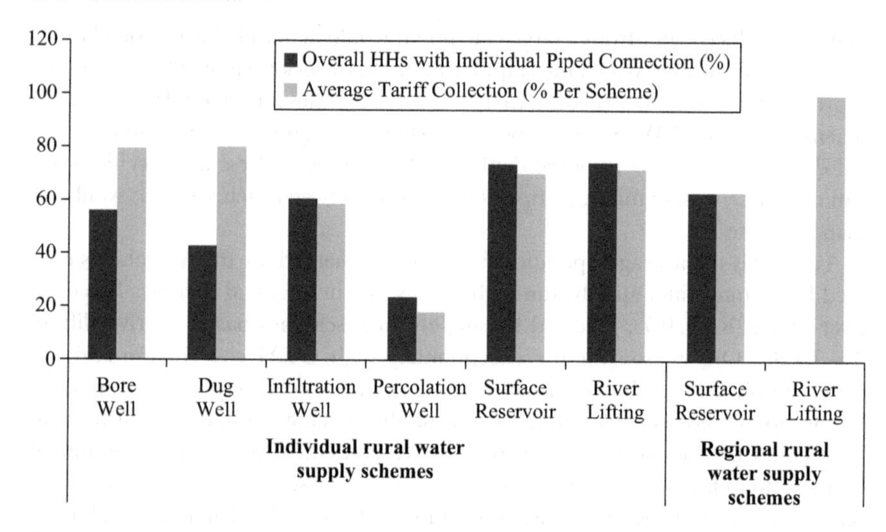

Figure 10.3 Scheme and Source-Wise Individual Household Connection and Tariff Collection

Source: Authors' own analysis using primary data collected from various water supply agencies

regional water supply schemes, average bulk tariff varied from Rs 6 per kiloliter to Rs 8.33 per kiloliter.

Overall, average tariff collection (tariff recovery as a percentage of the total demand) was significantly higher for schemes that tap water surface water (70.5%) than those that tap groundwater and subsurface water (63%). It is to be kept in mind that the surface water-based schemes also have significantly high proportions of the households under individual tap connection (Figure 10.3). Technology wise, average tariff collection was highest for regional schemes lifting river water (99.8%), which was followed by schemes extracting groundwater through dug wells (79.8%) and a bore well (79.25%); individual schemes tapping water through river lifting (71.7%) and surface reservoir (70%); regional schemes dependent on surface reservoir (63%); and schemes extracting subsurface water through water through infiltration (58.6%) and a percolation well (18%).

Degree of decentralization in the management of rural water supply

Under the sector reforms, VWSC is expected to perform many functions relating to the planning, design, execution and management of water supply at the local level for promoting decentralized management and community participation. For understanding the degree of decentralization in management, observed roles and responsibilities of GP/VWSC was compared with those that are expected.

As regards planning and designing of schemes, in 10 out of 12 selected schemes, planning and design of water supply schemes was undertaken by either technical wing of ZP or MJP. Only in two individual schemes, one based on a dug well and other one on a percolation well, planning and design of the scheme was carried out by VWSC. Further, in only 50% of the selected cases, GP built the scheme, but it was found to operate all the selected individual schemes. However, as per the approved administrative procedures, all single village piped schemes with project costs ranging from INR 5 million to above INR 50 million Rs have to be executed by GP. But MJP and ZP were found to have executed two of the selected 7 such individual schemes.

In the case of single-village schemes, in most cases, the scheme was built by the GP, while in one case, the scheme was built by ZP, and in another the scheme was built by the MJP. Nevertheless, in all cases, the scheme was operated by the GP. In the case of regional schemes, MJP emerged as the major player, performing the function of main system management, while the operation and maintenance of the scheme components at the village level was done by the VWSC. But that, too, is done under the technical supervision of MJP. It is quite clear from this arrangement that the majority of management functions are in the hands of agencies such as MJP, and the ZP or the GP do not take charge of the system operation.

Table 10.6 provides the details of the expected roles and number of VWSCs actually performing these roles.

Human resource capabilities

Although there are no standard norms being followed in the state regarding number of agency staff required to handle the water supply, distribution and maintenance functions, it was found that the number of staff handling such functions was extremely low. Overall, per 1,000 households covered by water supply, only 0.33 technical staff, 0.21 managerial staff, 0.14 financial staff, 1.27 other nontechnical staff, and 2.94 contract-based staff were found to be handling the water supply functions. The number of technical staff was too low considering that the major functions for executing projects in cases of individual-village schemes and almost all the functions in cases of regional water supply schemes are handled by them.

Among the schemes, the highest number of technical staff per 1,000 covered households was found in individual groundwater-based schemes (1.2) and the lowest in individual surface water-based schemes (no employee). Managerial staff was also highest in case of groundwater-based schemes (3.5 per 1,000 covered HHs) and lowest in individual surface water-based schemes (no employee). However, nontechnical staff per 1,000 covered households was highest in individual surface water-based schemes (4.3) and lowest in regional surface water-based schemes (1.1). Further, only surface water-based regional schemes had employed the contractual staff.

Governance of water supply

While governance is the art of rule making, the key areas for rule making that are relevant in the context of rural water supply and that are being addressed in the research pertain to who decides on the following: a) the type of scheme, b) the

Table 10.6 Roles Performed by the VWSC Against the Expectations

Functions Expected of VWSC	*Number of VWSCs Performing the Following Function*	
	Regional Scheme (Total of 183 VWSCs)	*Individual Schemes (Total of 7 VWSCs)*
Ensuring community participation and decision making in all phases of scheme activities	57	7
Organizing community contributions toward capital costs	0	7
Operating bank account for depositing community cash contributions and O&M funds	57	6
Preparation of village water security plan	0	3
Planning, designing and implementing all water-related activities in the village	57	7
Planning, designing and implementing all sanitation-related activities in the village	57	3
Procuring construction materials/goods and selection of contractors (where necessary) and supervision of construction activities	57	6
Ascertaining drinking water adequacy at the household level including cattle needs	48	4
Tariff collection	57	7
Empowering of women for day-to-day operation and repairs of the scheme	0	5
Participation in communication and development activities in other villages	9	2
Testing of supplied water quality	57	7
O&M supervision and monitoring	57	7

Source: Authors' own analysis using primary data

source of water for the scheme, c) the water supply schedules, d) duration and timing of water supply, e) the individual connection charges, f) mode of pricing water and water rates, g) the penalty for nonpayment, h) presence of complaint redressal mechanism and i) the frequency of water quality monitoring.

The results show that in the case of single-village schemes, the governance of rural water supply is more or less decentralized, with mostly the GP deciding on the type of scheme and source, water supply schedules, duration and timing of water supply, the individual connection charges, mode of pricing water and water charges, penalty for nonpayment of water charges wherever it exists and frequency of water quality monitoring. To an extent, the ZP is also found to be involved in performing some of the governance-related functions. Contrary to this, in the case of regional water supply schemes, the ZP and MJP together replace the GP/ZP, with some of the governance roles being performed only by the MJP. For instance, in the case of regional schemes based on reservoirs, the MJP was perceived to have a significant role in deciding on the type of scheme, type of source, water supply schedule, timing and duration of water supply, individual tap connection charges, mode of pricing water and water rates and frequency of water quality monitoring. At the same time, in the case of single-village schemes, most of these decisions are taken on primarily by the GP and, to an extent, by the ZP.

Summing up

Clearly, the rural water supply and sanitation sector in Maharashtra has shown impressive progress in terms of coverage in the state of Maharashtra. But to achieve the ultimate goal of source sustainability, water security, more effort is required. On the water supply front, large dependence on groundwater-based sources, on the premise that the local institutions such as the village panchayats could run and manage these schemes, has led to poor system performance.

A comparative assessment of performance of 12 rural water suppy schemes in six divisions of Maharashtra shows that the management performance of single-village schemes that are based on surface water is better than that of their groundwater and subsurface water counterparts. The overall performance of water supply schemes based on surface reservoirs is also better than that of schemes based on groundwater and subsurface water. Although the degree of decentraliztion and community participation in management of the scheme is comparitively better for the single-village schemes based on groundwater, this has come at the cost of sustainability of water supplies, not only from the point of view of providing sufficient quanties of water for meeting domestic water requirements throughout the year, but also in terms of the cost effectiveness. Further, groundwater availability is extremely limited in the state (GOM, 1999). The Groundwater Act has been largely ineffective in regulating groundwater use by irrigators for protection of drinking water sources.

Clearly, there is a tradeoff between 'sustainability' and 'decentralized governance and management.' Even the goal of decentralization in governance and management of water supply scheme hasn't been achieved. As per the performance assessment of selected individual water supply schemes, all the village water and sanitation committees were reported to be performing only six out of the 13 key roles vested in them, while the remaining roles are performed by a lesser number of VWSCs. The situation is even grimmer in the case of selected regional water supply schemes, where only in less than one-third of the cases were the VWSCs performing some of the key roles, while in the rest of the cases, they are totally dysfunctional.

Currently, the willingness on the part of the GP or ZPs to take charge of running the regional water supply schemes is largely absent. Most of the regional schemes are still run by the MJP, while in a very few cases where the number of villages covered is small, the ZP had taken over the system running. The important reasons for this are the technical sophistication of the schemes, the lack of qualified staff to take care of the maintenance and fear of the financial burden of 'high O&M costs.' Further, the response of the VWSC in terms of taking over the village-level maintenance of the scheme is also not encouraging, as they are at best equipped to mobilize the village community for social action. Therefore, the handing over of responsibilities to them, without building their institutional capacities from the technical know-how, management, finance and governance points of view, is leading to a situation of 'leapfrogging.' The reform agenda has missed these critical points.

In regions with similar physical characteristics vis-à-vis hydrology, geo-hydrology and topography, the regional water supply schemes and single-village schemes that are based on surface reservoirs ensure water supplies of higher dependability to the rural areas, as compared to schemes that are dependent on wells. The quality of water is better for schemes that tap water from surface reservoirs, as compared to those that tap water directly from rivers or groundwater. Over and above this, contrary to the popular notion that the O&M cost of regional water supply schemes is higher than that of decentralized individual village-based schemes that are dependent on local sources such as wells, our analysis shows that the O&M cost per household is lesser for single-village schemes based on surface reservoirs, and that of regional water supply schemes based on reservoirs is comparable with dug well-based single-village schemes. This is in addition to the advantage of higher dependability and longer life of the systems. More importantly, the factor that causes major differences in the O&M cost is the higher salary being paid to the staff of the water supply agency.

The above recommendation on going for schemes based on surface reservoirs is further strengthened by the review of water audit report of irrigation projects in Maharashtra, which shows that a large amount of water remains unutilized in irrigaiton reservoirs of the state at the end of the irrigation season. Surprisingly, the nonirrigation use from many schemes is less than planned use. During 2009–10, the total amount of unutilized water in the reservoirs was to the tune of 1,985 MCM. This is quite substantial if one considers the total amount of water required to meet domestic water needs in the entire state of Maharashtra. It is a very appalling that when thousands of villages including those in the water-rich regions of the state reel under water shortages for meeting survival needs, the water stored in such large and expensive infrastructure goes unutilized. Such a precarious situation prevails due to lack of infrastructre for transporting the water from these reservoirs to the areas experiencing drinking water shortages in different parts of the state.

Overall, the reform agenda seems to have gotten compromised on physical sustainability in rural water supply and the role of technology choice in ensuring that. Instead, it had focused simply on decentralization. Here again, institutional arrangements are not put into place to ensure that the local institutions responsible for decentralized management have adequate institutional capabilities for governance and management of rural water supply.

In view of the water scarcity experienced in many districts of the state, the aspects related to water supply quantity and water quality need to be tackled differently. For that, the approach of planning, designing and implementing rural water supply schemes as single-use systems (domestic use) has to get replaced by schemes that can take care of multiple needs (both domestic and productive use) of village communities. Further, if sustainability of the source has to become a priority, then the strategy has to change from groundwater-based individual supply schemes to reservoir-based regional supply schemes, unless proper enforcement of groundwater regulation is done. With the large number of major and medium reservoirs scattered across the state, and with around 2,000 MCM of water remaining unutilized in these storage systems at the end of the irrigation season every year, augmenting water supplies of the existing village or regional schemes should be possible with the building of a large distribution network using pipelines.

On the institutional development front, the VWSC needs to be properly trained so that it can take effective part in design and O&M of multiple-use RWSS schemes. Thus, communities can feel the incentives and take active part in operation and maintenance of the systems. The scheme with multiple uses should consider all the available sources in the area in order to judiciously use the available resources. In the case of regional water supply schemes covering large numbers of villages, the system operation should be handled by the MJP. As it is involved in planning and execution, such an approach would improve the operational efficiency. The financial and human resource capacities of the MJP need to be strengthened so that it can play an effective role in rural water supply management.

Note

1. Overdependence on groundwater-based schemes resulted in drying up of the sources such as open wells, tube wells and hand pumps as a result of overexploitation of groundwater and drops in water levels.

References

Cullet, P. (2011). Realisation of the Fundamental Right to Water in Rural Areas: Implications of the Evolving Policy Framework for Drinking Water. *Economic and Political Weekly, 46*(12), 56–62.

Das, K. (2006). Drinking Water and Sanitation in Rural Maharashtra: A Review of Policy Initiatives. Ahmedabad: Gujarat Institute of Development Research.

Gawade, V. (2012). *Maharashtra Sector Status Report: Water Supply*. Draft Report, Water and Sanitation Program-World Bank, Government of Maharashtra.

Government of Maharashtra (GOM) (1993). The Maharashtra Groundwater (Regulation for Drinking Water Purposes) Act, 1993. Maharashtra: Author.

Government of Maharashtra (GOM) (1999). Maharashtra Water and Irrigation Commission Report. Maharashtra: Author.

Government of Maharashtra (GOM) (2003). The Maharashtra State Water Policy (2003). Maharashtra: Author.

Government of Maharashtra (GOM) (2009). Maharashtra Groundwater (Development and Management) Act, 2009. Maharashtra: Government of Maharashtra.

Pattanayak, S. K., Poulos, C., Wendland, K. M., Patil, S. R., Yang, J. C., Kwok, R. K. and Corey, C. G. (2007). *Informing the Water and Sanitation Sector Policy: Case Study of an Impact Evaluation Study of Water Supply, Sanitation and Hygiene Interventions in Rural Maharashtra, India*. Working Paper 06_04., Research Triangle Park, NC: Research Triangle Institute International.

Phansalkar, S. and Kher, V. (2006). A Decade of the Maharashtra Groundwater Legislation: Analysis of the Implementation Process. *Law, Environment and Development Journal, 2*(1), 69–83.

Reform Support and Project Management Unit (RSPMU) (2004, January–March). Maharashtra Rural Water Supply and Sanitation Project – Jalswarajya. Quarterly Progress Report. Maharashtra: Reform Support and Project Management Unit, Water Supply and Sanitation Department, Government of Maharashtra.

Sampat, P. (2007). Swajaldhara or 'Pay'-jal-dhara: Sector Reform and the Right to Drinking Water in Rajasthan and Maharashtra. *Law, Environment and Development Journal, 3*(2), 3–125.

Sangameswaran, P. (2010). Rural Drinking Water Reforms in Maharashtra: The Role of Neoliberalism. *Economic and Political Weekly, 45*(4), 62–69.

Water and Sanitation Programme (WSP) (2004). *Alternate Management Approaches for Village Water Supply Systems Focus on Maharashtra*. New Delhi: Water and Sanitation Programme-South Asia.

World Bank (2008). *Review of Effectiveness of Rural Water Supply Schemes in India*. New Delhi: The World Bank.

World Bank (2011). *Jalswarajya Project: Easing the Burden of Water for Villagers in Rural Maharashtra*. New Delhi: The World Bank.

World Bank (2012). *Draft Report on Maharashtra Sector Status: Water Supply*. Maharashtra: Government of Maharashtra, Water and Sanitation Program-World Bank.

11 Future impacts of agri-business corporations on global food and water security

M. Dinesh Kumar

Introduction

The concept and practice of virtual water trade and activities of agri-business corporations have caught the attention of academia around the world for their implications for global food and water security (Cousin, 2012; Kumar, 2012; Sojamo et al., 2012). In this chapter, we will first discuss the recent trends in global agri-business wherein the dominance of western agri-business corporations in international food trade are being challenged by China and some oil-rich countries of the Middle East. Subsequently, an examination is made to find whether this can be treated as an attempt by the rising southeastern economic powers to establish hegemony over global virtual water trade and whether this has any significance for the global water security. This is done in the light of the received wisdom about the real determinants of what governs global virtual water trade and the emerging scenario with regard to changing global land-use patterns. Particularly, critical examination is made on the claim regarding the growing foreign direct investment on land in Africa by the emerging Asian and Arab economies; and their increased competition over the sources of global food supplies is a strategy to challenge the western hegemony over 'international virtual water flow.' We further examine whether the ongoing power shift in the global food market from the industrialized West to the industrializing East would provide any new governance architecture for water security and stewardship of water resources.

Virtual water trade and global water security

Fair distribution of freshwater between humans and nature, and among all sectors of use, is one of the main challenges of the 21st century (Zehnder and Reller, 2002). Given the fact that vast differences exist in natural endowment of water resources across the globe, the idea of redistribution of water and its scarcity has caught the imagination of many researchers dealing with water management in the face of the growing regional water shortages. Since Professor Allan's pioneering work in 1993, a few research-based articles were published in academic journals on virtual water trade. While some questioned the utility of the very concept itself (Wichelns, 2001, 2010), some of them looked at whether virtual water could be used as an instrument for tackling water scarcity problems, globally or regionally (Kumar and Singh, 2005; Singh, 2004).

The very fact that food production requires much more water than do drinking and hygiene (Zehnder and Reller, 2002) means trading in agricultural commodities is increasingly being viewed by researchers as a useful mechanism for redistribution of water in large quantities and saving water in the country of import (see, for instance, Delgado et al., 2003; Parveen and Faisal, n.d.; Sojamo et al., 2012).

During the past 2 decades, the theory of virtual water has been advanced by many scholars. Hoekstra (2003) suggested concept of 'water footprint,' which indicates the virtual water trade balance of a country that could become a useful indicator of a nation's dependence on water in the global system and is also becoming essential for developing a more rational country policy with respect to virtual water trade. The 'water footprint' was calculated based on the use of domestic water resources and the net import of virtual water.

Wichelns (2001, 2003) emphasized that including virtual water import as a policy option for food and water security should involve considerations such as providing for national security, promoting economic growth, creating employment for people and reducing poverty, so as to minimize its impact on local economy, culture, employment and poverty. Mori (2003) takes a step forward and argues that a set of international norms, principles, rules and decision-making systems are to be designed and successfully converged upon to prevent virtual water trade from leading to even more conflicting situations in the rapidly changing global trading system.

The recent past has witnessed methodological advancement in the analysis of virtual water globally as well as within regions. Hoekstra and Hung (2003) quantified the total virtual water trade globally. For this, the authors used a basic approach of multiplying international crop trade (ton/year) by its associated virtual water content (m^3/ton). The total water use by crops in the world has been estimated by Rockström and Gordon (2001).

Chapagain and Hoekstra (2003) for the first time developed a methodology and quantified the virtual water flows between nations through trading of livestock and livestock products. The results were then combined with the estimates of virtual water trade associated with international crop trade as reported in Hoekstra and Hung (2002, 2003) to get a comprehensive picture of total virtual water trade in the agricultural sector. Allan (2003) showed that the Middle East was very poorly endowed with freshwater. In 2000, the Middle East and north Africa were importing 50 million tons of grain annually, satisfying the largest demand for water in the region for food production.

Wichelns (2003) showed that food security at the national level in Egypt is achieved through a combination of domestic production and import of agricultural products. Increase in domestic food and fodder production also may have been essential in improving incomes, reducing poverty and enhancing household food security in rural area. Haddadin (2003) developed a methodology to measure the water embedded in the imports of food commodities. Taking Jordan as an example, he argued that exogenous water accounts for a substantial portion of the water available to arid and semi-arid countries.

El-Fadel and Maroun (2003), by taking the case of Lebanon, argued that local policy constraints may cause hindrances to countries in utilizing their potential to produce surplus food, and water supply shortages will result in the country ending up with food imports. Earle and Turton (2003), who studied virtual water trade amongst countries of the Southern African Development Community (SADC), found that while as a group, the SADC countries are well endowed in water resources, they display large-scale variations in the temporal and spatial distribution of rainfall and water resources, making averages more or less meaningless. Their study developed a typology of factors influencing the degree of reliance on virtual water by states in general.

Meissner (2003) analyzed virtual water trade in SADC member states from a political, economic and environmental perspective. He argued that for many of those countries that are experiencing food insecurity problems and water stress, pursuing the virtual water route would be a sound strategy because of its economic and political benefits.

Globally, research studies dealing with virtual water have largely viewed food trade as a tool for mitigating the stress caused by scarcity of water needed for food production (see, for instance, Allan, 1997; Hoekstra, 1998; Parveen and Faisal, n.d.; Earle and Turton, 2003; Meissner, 2003). Nevertheless, in the recent years, several scholars have questioned the idea of virtual water trade as a tool for addressing water scarcity not only on practical grounds (Kumar and Singh, 2005) but also on conceptual (Merrett, 2007), theoretical (Cousin, 2012; Wichelns, 2004, 2010) grounds.

Warner (2003) argued that virtual water analysis should consider non-water factors as trade in food and other water containing products as not only concerns virtual water, but also as virtual land, virtual labor, etc. They also viewed food insecurity problems experienced by many countries as a problem associated with water shortage or lack of technologies to utilize it, while El-Fadel and Maroun (2003) attributed it to local policy factors also.

As argued by Kumar and Singh (2005), there are water-management objectives that virtual water transfer does not address, the most important of which is water productivity. Even for crops that have high water use efficiency, water productivity could be much lower, resulting from inefficient use of irrigation water. Unfortunately, the distinction between water use efficiency and water productivity is often not explicitly stated in irrigation-management literature (Grismer, 2001). This results in little attention being paid to the mechanisms for improving actual field efficiency in water use. While over the past 25 years, there have been significant improvements in efficiency of water use in irrigation globally, leading to leveling off of water use (Gleick, 1998), it has mostly been in developed countries (Sorooshian et al., 2002).

Merrett (1997) was one of the first economists to criticize the concept of virtual water as a mere metaphor. He argued that the water content contained in the products is much less than the quantity that was used in production, so after all, there is no such thing as virtual water trade – it is simply trade of food. Wichelns (2004) argued that there is inconsistency between the comparative advantage theory and virtual water trade. He argues that the virtual water concept addresses neither production technologies nor the opportunity costs. Wichelns (2010) extended this argument that the issue of opportunity costs of water resources should be taken into account while determining its optimal allocation.[1] Nevertheless, this argument failed to take into account the role arable land has in allocating water resources for crop production.

A study done by Dabrowski et al. (2008) elaborates on the concept of crop productivity by looking into the case of South Africa. While South Africa is one of the many nations that suffer from water scarcity, it is the largest producer and exporter of maize. Evidence shows that among the SADC region, South Africa has the highest crop productivity, combining both green and blue water resources. Therefore, once again, water endowment on its own cannot be taken as a decision maker for food production and trade of agricultural products. Guan and Hubeck (2007) also make an argument related to the issue of opportunity costs. They conducted a case study on China and observe that northern China, which is water scarce, is actually a net exporter of water-intensive products, while southern China, a water-abundant region, imports water-intensive products. They argued that water price, labor availability, soil and land quality are the driving force of this phenomena.

In a nutshell, research carried out till now only looked at how water endowment, soil quality, climate, labor and agricultural trade policies could influence virtual water trade patterns.

Drivers of global agri-business: land scarcity versus water scarcity

Internationally, discussions on global water scarcity are rich and plenty. This was particularly from the point of finding sufficient amounts of water to grow food for the growing population, as the largest demand for water consumption comes from agriculture, particularly food production. Food security problems are often falsely linked to water scarcity. Perhaps because of this reason, the movement of agricultural commodities across nations and continents has been mostly viewed through the lens of virtual water, as an attempt to gain control over water. But many water-scarce regions of the world are already producing surplus food for export to water-rich regions, be it the western United States, southern Spain, north China or northwestern Punjab. In these regions, water scarcity is felt as a result of excessive requirements of water for crop production induced by the presence of large amounts of arable land, and the demand for cereals and other agricultural commodities from water-rich regions, which do not achieve food self-sufficiency. Yet there has not been any attempt to link international trade in food to land scarcity.

Kumar and Singh (2005) further found, based on their analysis of data on virtual water trade from 131 countries, that at the operational level, virtual water trade across countries does not result in 'global water use efficiency' improvements or 'distribution of scarcity' due to the reason that at the global level, virtual water flows out of water-scarce countries that are land-rich, to water-rich countries that are land-scarce. They argued that water richness does not guarantee food self-sufficiency, but access to arable land does. Many countries indulged in food import not because of shortage of water, but because of limited availability of arable land to cultivate crops, and often what is being achieved through food trade is an improvement in global land-use efficiency (Kumar and Singh, 2005). Verma et al. (2008) analyzed the virtual water flow data for 15 provinces (states) of India and corroborated the findings of Kumar and Singh (2005) with respect to the role of arable land (gross cropped area) in determining the virtual water trade balance/ deficit.

The problem of land scarcity is likely to grow in an unprecedented manner in the coming years, posing a major threat to global food security. There are three major reasons for this. First, many developed countries, including those that export agricultural commodities, are likely to free a portion of their crop land for ecological uses, such as reforestation and conservation of wetlands and pastures. This trend is also being witnessed in some of the fast-growing developing economies, which are also the largest producers of agricultural commodities. This is done through increased reliance on imported food and wood products, creation of off-farm jobs, foreign capital investments and remittances. This forms part of their effort to transfer the environmental cost of increased consumption of goods to other territories (Lambin and Meyfroidt, 2011), and is the result of the growing demand for improving environmental management services in these countries, owing to rising per capita incomes. Two such countries are Brazil and South Africa. As Lambin and Meyfroidt (2011) note, such land-use zoning decisions made by these countries may trigger compensating changes in trade flows and thus indirectly affect land-use changes in other countries.

Second, with the rapidly growing domestic consumption owing to rising income levels and with large amounts of foreign exchange being spent for import of agricultural commodities – particularly those that are at the higher end of the food chain such as chicken, beef and dairy products – at least a few countries, which have reached the zenith in terms of expanding crop land, are more likely to look for new sources of food supplies under fear of future food security problems (Cousin, 2012). The recent move by China to purchase cropland in sub-Saharan Africa is one such attempt. Third, corporations in some of the oil-rich countries such as Saudi Arabia, which are not endowed with good arable land, are likely to look for land outside their territory for raising crops for imports as a major business proposition. As such food systems without much dependence on foreign countries are likely to give strategic advantage to their national governments in terms of food security, these corporations are likely to receive large incentives from their national governments.

Having described the global trend with respect to land use, the next step is to examine where the pressure on land resources for crop production is likely to increase to make up for the increase in global demand. Several factors would determine this. The availability of arable land, existing land-use intensity, the availability of cheap labor, market price of land and access to production technologies are the most important among them. The land-use intensities are already very high in Asian countries, which already have very high population density and hence are unlikely to witness further increase in land-use intensity. Sub-Saharan Africa, on the other hand, has a very low percentage of its arable land under cultivation and also has very low crop duration index (Siebert et al., 2010), and a much smaller percentage of this land is irrigated.[2]

Major results of the work of Siebert and others (2010) are presented in graphical form in Figure 11.1. It shows that cropping intensity (measured as the ratio of the total area harvested as a percentage of the cultivable land) is 0.45 (southern Africa) to 0.83 (western Africa) in African regions, whereas the crop duration index is in the range of 0.27 (southern Africa) to 0.47 (western Africa) in these regions. The corresponding values for South Asia were 1.00 and 0.50.

Land and labor are also cheap in these countries. It is a widely known hypothesis that agricultural intensification will spare land for nature. The underlying premise in this argument is that with a fixed demand for agricultural commodities, higher yields decrease the area that needs to be cultivated. But in reality this may not happen. In the short run, with more intensified agriculture using efficient production systems, the price decline that can occur as a result of the decline in cost of production can generate a demand increase depending on the price elasticity of demand for agricultural commodities. This can lead to increase in cultivated area (Angelsen and Kaimowitz, 2001). In the long run, the expansion depends on the impact of technological progress on economic growth and population rise (Lambin and Meyfroidt, 2011).

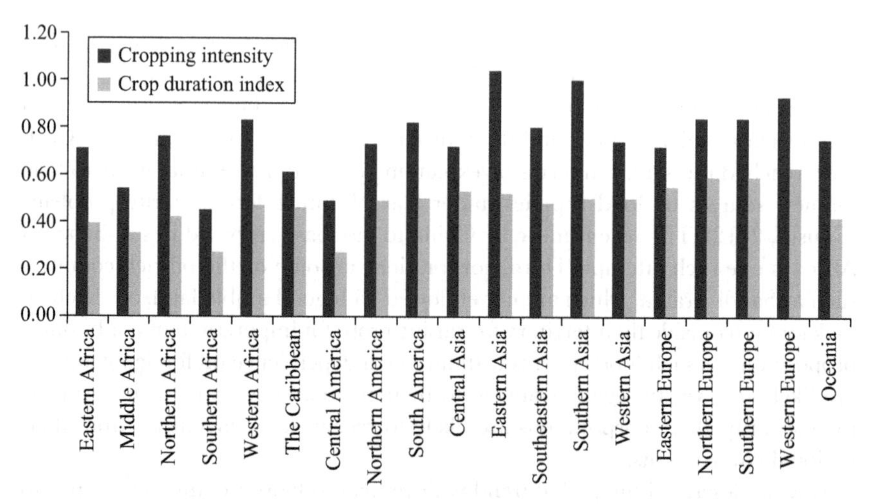

Figure 11.1 Cropping Intensity and Crop Duration in Different Regions of the World

Source: Based on Siebert et al., 2010, Table 1, p. 1630

The operations of these agri-business corporations, which purchase large tracts of land from the original title owners in these countries, could limit the ability of the native farming communities to own arable land in future, owing to a sudden rise in price of agricultural land. The ownership of land titles could give these corporations effective control over the water resources through irrigation water rights or entitlements. On the one hand, the poor rural communities of these countries have limited capacity to adopt modern agricultural technologies and irrigation facilities due to poverty, and therefore obtain poor yields in their farms. Sub-Saharan Africa houses some of the most food insecure nations of the world, with several millions of people suffering from malnutrition and very high rates of infant mortality (IFPRI, 2011; Wiesmann, 2006). Reductions in size of operational holdings over time could further threaten their subsistence by reducing their crop outputs and farm surplus, although those who sell their land to these corporations would make significant monetary gains in the short run. On the other hand, control over factors of production could give monopoly power to large agri-business corporations in deciding the food prices, affecting the regional food security.

Agri-business corporations and global water security

Globally, there has been a growing concern over the environmental consequences of the increasing control that large agri-business corporations exercise over food systems. Big agri-business firms today decide how America produces its food (Dutzik et al., 2010).[3] Sojamo et al. (2012) argued that with the recent global food crises, the agro-food systems tend to be subject to manipulations by powerful agri-business players, with consequences for global water security. According to them, the Western countries have dominance over the virtual water embedded in international trade in agricultural commodities, which they called 'virtual water hegemony.' They further argued that the foreign direct investment in land by emerging Asian and Arab economies and their increased competition over the sources of global food supply were strategies to challenge the 'virtual water hegemony' of the Western agri-business. The work by Sojamo et al. (2012) is unique in terms of how the political economy of global agri-business and food trade is changing with the rising economies of the East taking over the more rooted, western agri-business corporations.

But their analysis failed to take into account the analyses that looked at the key drivers and determinants of global 'virtual water' trade, and therefore suffered from several flaws. For instance, Kumar and Singh (2005) in their analysis showed that the most important determinant of virtual water flow globally is the access to arable land (expressed in terms of per capita gross cropped area). The reason was that with the increase in per capita gross cropped area, the per capita 'effective water withdrawal in agriculture'[4] increased in a linear fashion. In spite of the fact that irrigation is not a significant contributor in expanding cropland for most countries, and that there are major variations in green water use across countries (from a low of 1.7% to 50% of the effective water withdrawal for agriculture), such a strong relationship exists. The degree of use of soil water and irrigation water were found to be inverse functions of the arable land availability, as an inverse relationship exists between rainfall (water richness) and arable land availability per capita (Kumar and Singh, 2005).

More importantly, this research showed that virtual water export mostly took place from relatively water-scarce countries, which have plenty of arable land, to water-rich countries, which are poorly endowed in arable land. Indonesia, Bangladesh, the Philippines, Malaysia, Nepal and the United Kingdom have significant virtual water trade deficits. They are relatively water rich – some of them extremely water rich. Further, the study noted on the basis of empirical analysis that the per capita annual renewable water resources of a country is not a determinant of virtual water trade globally, the reason being that their water richness resulted from high precipitation and runoff and recharge, but lacked sufficient amounts of arable land, which could use the water for crop production. Many water-scarce countries that are rich in arable land resources are able to tap water in the soil profile (green water) and exploit renewable and nonrenewable water resources to produce crops, which is normally considered in the estimation of renewable annual water resources (Kumar and Singh, 2005).

Sojamo et al. (2012) used the example of a Saudi Arabian business tycoon to illustrate the point that investment in land is a strategy of economically aspiring water-insecure countries to challenge the hegemonic agri-food systems of the Western agri-business corporations (Sojamo et al., 2012, p. 178). But this is an ill-conceived argument. The reason is that although Saudi Arabia is an extremely water-scarce country, it does not stand as a great example of water insecurity, but for acute shortage of productive soil, with a remarkable portion of its land as deserts. The reason is that the country's economy is not dependent on production of crops, but on oil reserves, with the demand for water mainly for rural and urban domestic uses.

While there are many regions in the world that are even more water rich, requiring much less investment for irrigation development, the Saudis' interest in Ethiopia is quite comprehensible. The region has a large amount of arable land, which is available at a low price (of $12 USD per ha) (Sojamo et al., 2012), and a remarkable amount of unutilized surface water resources, with plenty of rural labor force. Kumar and Singh (2005) theoretically argued that the availability of arable land and water resources provide the most ideal condition for production of surplus agricultural commodities for trade. This was supported by Cousin (2012), who involved empirical work involving data from 167 countries on net per capita virtual water export and gross cropped areas and renewable water resources. Water richness of an area alone does not guarantee the ability to generate agricultural surpluses, when land is very scarce (Kumar and Singh, 2005). There are in fact, many water-rich regions and countries in the world, which in spite of having access to modern crop technologies and irrigation facilities, are not able to become food self-sufficient. A few examples are Indonesia, Bangladesh and Japan. Australia, the largest exporter of agricultural commodities (virtual water) in per capita terms, is able to achieve this unique distinction not because of high precipitation and water richness, but because of the access to enormous amount of arable land, i.e., 24 ha per capita (Kumar and Singh, 2005).

In fact, large parts of Australia are drought prone and extremely water scarce – in terms of precipitation they receive. But the country could convert this unfavorable context into a great opportunity to become the largest exporter of virtual water in per capita terms (Kumar and Singh, 2005), by producing surplus cereals and other agricultural crops by using the vast tracts of cultivable land the country is endowed with, with a relatively small population to support. For this, it fully exploited the renewable and nonrenewable water resources (groundwater stock in aquifer basins like the Great Artesian Basin), the other excellent factors of production and favorable irrigation and agricultural trade policies.

Several writers in the recent past argued that India and China would be the first to bear the brunt of water scarcity and that they would become net importers of food (Sojamo et al., 2012). For instance, Sojamo et al. (2012) argued that water scarcity will become a great obstacle to agricultural development in China, India and other Asian countries. This is far from the truth. The fact is that these countries are not homogeneous in terms of their water-resource endowments. They have both water-rich and water-scarce regions (Kumar and Singh, 2005; Yang, 2002). Many water-scarce regions of China, India and Pakistan have picked up high agricultural growth trajectories, and are already net exporters of food and agricultural commodities, be it the north China plains, northwestern India or the south Indian peninsula or Pakistan Punjab. Some of these regions have a lot of arable land that is still under rain-fed production, whereas, the water-abundant regions of India (eastern Indian provinces of Bihar, West Bengal, Assam and Uttar Pradesh) and China (south China) are net importers of food grains from the water-scarce regions of the respective countries. One of the major constraints in enhancing agricultural productivity in these water-rich regions is the extremely limited access to arable land (Kumar, 2003; Kumar et al., 2012). Hence, future food insecurity problems could occur because of acute shortage of arable land resources (Kumar, 2012). This argument was supported by Cousin (2012), who argued that countries such as China are securing agricultural land area in different countries due to the concern of future food security.

Therefore, the contention by some authors (for instance, Sojamo et al., 2012) that those countries would become vulnerable to food insecurity problems due to water scarcity seems to be questionable. Such countries have the option of reducing the imbalance in land and water resources through internal redistribution of water using large water transfer infrastructure,[5] rather than trading in virtual water (read as food and other agricultural commodities), internationally.

Further, the observation by Sojamo et al. (2012) that international trade can contribute to water-use efficiency and global water security if commodities are traded from areas of high water productivity to areas of low water productivity resulting in global water-use efficiency does not work in reality. Most of the agriculturally prosperous regions with high land productivity (ton/ha) are in the semi-arid and arid tropics, owing to the climatic advantages. They also have the advantage of low population density, with fewer mouths to feed. Egypt, Pakistan and Indian Punjab are examples of high wheat yields, for instance. But many of the low productivity (ton/ha) regions are water rich, and face land scarcity. The best example is Bangladesh. It is also burdened by high population density. The net effect is that a lot of virtual water gets exported from water-scarce regions to water-rich regions through the export of agricultural commodities. Obviously, this cannot be treated as a net gain in water for the importing countries, but rather in virtual land and labor.

Though the countries in temperate climates, which also export crop outputs (virtual water) from North America and western Europe, have low blue water use in crop production, their green water component is also less because of the cold climate. This is contrary to the argument by Sojamo et al. (2012). Ultimately, what determines the trade surplus or deficit in virtual water flow is the gross area under production of the crops that are being traded and the climate under which they are produced, which the authors conveniently ignored. Even with highly water-efficient crops, the total effective agricultural water use in a water-scarce region could be much higher than that in a water-rich region due to intensive land use for crop production, which depends on irrigation (Kumar and Singh, 2005). As noted by Singh (2004), while many crops in semi-arid north Gujarat are low water consuming, intensive production of milk in the region – estimated to take 3.2 m^3 of water for a liter of buffalo milk and 2.5 m^3 of water for a liter of crossbreed cow milk – a large chunk of which is exported to other regions in Gujarat, led to virtual water exports from the region (Singh, 2004). Meanwhile, possibilities of intensive land use through short-duration crops are very little in countries falling in high rainfall regions due to factors relating to agro-ecology and socio-ecology (Kumar and Singh, 2005). Cherry picking of a few food and oil seed crops from selected regions for analyzing virtual water trade flow provides a distorted picture.

In sum, the contention that the ongoing power shift in the global food market from the industrialized West to the industrializing East would provide new governance architecture for water security and stewardship of water resources is misplaced. The decisions of agri-business corporations for investment in land and water resources for crop production and dairying would be largely influenced by considerations of availability of cheap land, low investment requirements in provision of irrigation water, cheap labor and the geographic positioning of such regions vis-à-vis the export markets and their political dispensation, rather than their relative 'water richness' or 'water scarcity.' If these conditions are to be met for providing a favorable investment climate, it is quite obvious that the future food exporting regions may not necessarily be water rich, and the importing regions may not necessarily be water scarce. In that respect, it is quite likely that in future, the Middle East and north Africa, which even today are limited players in global agricultural commodity trade (virtual water trade), will only have a negligible role in deciding the global food trade and therefore virtual water flows. But, sub-Saharan Africa could become a major source of future food supplies for many developed countries and emerging economies.

Conclusions

As analyses by Kumar and Singh (2005) suggested and were subsequently corroborated by Cousin (2012), assessing the future food security challenges posed to nations purely from a water resource perspective provides a distorted view of the food security scenario, and is dangerous. Many water-rich regions may have to depend on food import due to land scarcity. On the other hand, assessing the water management challenges posed by nations purely from the point of view of renewable water availability and aggregate demands will also be dangerous. Access to water in the soil profile would be an important determinant of effective water availability for food production, which is a major source of water demand (Kumar and Singh, 2005). This should add a potentially new dimension to the global water scarcity debate. In a similar manner, it is incorrect to translate the food demand of a country or region into an equivalent water demand, if the same does not have sufficient arable land to produce that food. Through food import, what we probably gain is global land-use efficiency.

With globalization of economy, and with the pressure on many developed and some of the developing economies to free part of their cropland for ecological uses such as forest conservation, these countries are likely to witness land-use zoning decisions, causing major land-use changes. It is quite evident that under these circumstances, the agri-business corporations from the developed countries turn to regions where large amounts of arable land lies uncultivated, to gain control over international food trade, while those from oil-rich countries such as Saudi Arabia, with incentives from their national governments, might pursue the agenda of purchasing vast tracts of land in these countries not only to establish hegemony over international food trade, but also to ensure their own country's food security. On the one hand, the agricultural intensification and consequent expansion of cultivated area under corporate farming would severely affect the ability of the native communities in these regions to own farmland, thereby affecting their crop outputs and farm incomes. On the other hand, control over factors of production such as land and water would monopolize power to the corporations in deciding food prices. The combined effect would be an adverse impact on regional food security and livelihood security of farmers.

Notes

1. For example, the opportunity cost of producing rice in humid regions might be smaller than in arid regions. He supports his argument by presenting an example where, in the first case, two countries have the same water endowments with different productivity levels and, in the second case, they have different levels of water endowments. Even though the opportunity cost of producing one product in terms of the other remain the same, if production decisions are determined based on the hypothesis of virtual water trade, the country with a lower level of water endowment would change its production accordingly, while if we respect the comparative advantage theory, no change is required, as the opportunity costs remain constant.

2. A small percentage of the region's water resources are yet to be tapped (Falkenmark and Rockström, 2004), and the investment is water resources. Yet the region hasn't seen significant investments in irrigation development (Rosegrant et al., 2001).
3. Since 1993, the share of the nation's milk cows on large farms of 200 cows or more increased from 31% to 67%. Similar shifts toward concentrated animal feeding operations (CAFOs) have taken place in the chicken and pork industries (Dutzik et al., 2010).
4. The 'effective agricultural water withdrawal' was defined by Kumar and Singh (2005) as the sum of the irrigation water withdrawal and the green water use for crop production.
5. Such infrastructure could take water from water-rich basins, which have limited land resources, to water-scarce basins, which have sufficient amounts of extra arable land for irrigated production.

References

Allan, J. A. (1993). Fortunately There Are Substitutes for Water Otherwise our Hydro-Political Futures Would Be Impossible. In ODA (Ed.), *Priorities for Water Resources Allocation and Management* (pp. 13–26). London: ODA.

Allan, J. A. (1997). *Virtual Water: A Long Term Solution for Water Short Middle Eastern Economies?* Paper presented at the 1997 British Association Festival of Science, University of Leeds, September 9.

Allan, J. A. (2003). Virtual Water Eliminates Water Wars? A Case Study From the Middle East. In A. Y. Hoekstra (Ed.), *Virtual Water Trade: Proceedings of the International Expert Meeting on Virtual Water Trade.* Value of Water Research Report Series #12.

Angelsen, A. and Kaimowitz, D. (2001). *Agricultural Technologies and Tropical Deforestation.* Wallingford, United Kingdom: CAB International.

Chapagain, A. K. and Hoekstra, A. Y. (2003). Virtual water trade: A quantification of virtual water flows between nations in relation to international trade of livestock and livestock products. In: A. Y. Hoekstra (ed.), Virtual Water Trade: Proceedings of the International Expert Meeting on Virtual Water Trade. Value of Water Research Report Series No. 12. Delft, The Netherlands: UNESCO-IHE Institute for Water Education.

Cousin, E. (2012, June 12). Theoretical and Empirical Evaluation of Virtual Water Trade: The Significance of Arable Land versus Water Endowment. Maison des Sciences Economiques: Université Paris.

Dabrowski, J. M., Masekoameng, E. and Ashton P. J. (2008). Analysis of Virtual Water Flows Associated With the Trade of Maize in the SADC Region: Important Scale. *Hydrology and Earth System Sciences, 5*, 2727–2757.

Delgado, C., Rosegrant, M., Steinfeld, H., Ehui, S. and Courboi, C. (2003). *Livestock to 2020: The Next Food Revolution.* Food, Agriculture and Environment Discussion Paper 28. Washington, DC: International Food Policy Research Institute.

Dutzik, T., Madsen, T., and Ridlington, E. (2010). *Corporate Agribusiness and America's Waterways: The Role of America's Biggest Agribusiness Companies in the Pollution of our Rivers, Lakes and Costal Water.* Boston, MA: Environment America & Policy Center.

Earle A. and Turton, A. (2003). The Virtual Water Trade Amongst Countries of the SADC. In A. Y. Hoekstra (Ed.), *Virtual Water Trade: Proceedings of the International Expert Meeting on Virtual Water Trade.* Value of Water Research Report Series #12.

El-Fadel, M. and Maroun, R. (2003). The Concept of Virtual Water and Its Applicability in Lebanon. In A. Y. Hoekstra (Ed.), *Virtual Water Trade: Proceedings of the International Expert Meeting on Virtual Water Trade.* Value of Water Research Report Series #12.

Falkenmark, M. and Rockström, J. (2004). *Balancing Water for Humans and Nature: The New Approach in Eco-Hydrology.* New York: Earthscan.

Gleick, P.H. (1998). *The World's Water (1998–1999).* Washington, DC: Island Press.

Grismer, E.M. (2001). Regional Cotton Lint Yield and Water Value in Arizona and California. *Agricultural Water Management, 1710*, 1–16.

Guan D. and Hubek, K. (2007). Assessment of Regional Trade and Virtual Water Woes in China. *Ecological Economics, 61*(1), 159–170

Haddadin, M.J. (2003). Exogenous Water: A Conduit to Globalization of Water Resources. In A. Y. Hoekstra (Ed.), *Virtual Water Trade: Proceedings of the International Expert Meeting on Virtual Water Trade.* Value of Water Research Report Series #12.

Hoekstra, A. Y. (1998). *Perspectives on Water: An Integrated Model-Based Exploration of the Future.* Utrecht, the Netherlands: International Books.

Hoekstra, A. Y. (2003). Virtual Water: An Introduction. In A. Y. Hoekstra (Ed.), *Virtual Water Trade: Proceedings of the International Expert Meeting on Virtual Water Trade.* Value of Water Research Report Series #12.

Hoekstra, A. Y. and Hung, P. Q. (2002).*Virtual Water Trade: A Quantification of Virtual Water Flows Between Nations in Relation to International Crop Trade.* Value of Water Research Report Series No.11. Delft, the Netherlands: IHE.

Hoekstra, A. Y. and Hung, P. Q. (2003). Virtual Water Trade: A Quantification of Virtual Water Flows Between Nations in Relation to International Crop Trade. In A. Y. Hoekstra (Ed.), *Virtual Water Trade: Proceedings of the International Expert Meeting on Virtual Water Trade.* Value of Water Research Report Series #12.

International Food Policy Research Institute (2011). Global Hunger Index: The Challenge of Hunger-Taming Price Spikes and Excessive Food Price Volatility. Washington, DC: International Food Policy Research Institute, Concern Worldwide and Welthungerhilfe.

Kumar, M. D. (2003). *Food Security and Sustainable Agriculture in India: The Water Management Challenge.* Working Paper #60. Colombo, Sri Lanka: International Water Management Institute.

Kumar, M. D. (2012). Does Corporate Agribusiness Have a Positive Role in Global Food and Water Security? *Water International, 37*(3), 41–45.

Kumar, M. D. and Singh, O. P. (2005). Virtual Water in Global Food and Water Policy Making: Is There A Need for Rethinking? *Water Resources Management, 19*, 759–789.

Kumar, M. D., Sivamohan, M. V. K. and Bassi, N. (2012). *Water Management, Food Security and Sustainable Agriculture in Developing Economies.* Oxon, United Kingdom: Routledge.

Lambin, E. and Meyfroidt, E. (2011). Global Land Use Change, Economic Globalization, and the Looming Land Scarcity. Special Series of Inaugural Articles From the Members of the National Academy of Sciences.

Meissner, R. (2003). Regional Food Security and Virtual Water: Some Environmental, Political, and Economic Considerations. In A. Y. Hoekstra (Ed.), *Virtual Water Trade: Proceedings of the International Expert Meeting on Virtual Water Trade.* Value of Water Research Report Series #12.

Merrett, S. (1997). *Introduction to the Economics of Water Resources: An International Perspective.* Lanham, MD: Rowan & Littlefield Publishers.

Merrett, S. (2007). The Thames Catchment: A River Basin at the Tipping Point. *Water Policy, 9*(4), 393–404.

Mori, K. (2003). Virtual Water Trade in Global Governance. In A. Y. Hoekstra (Ed.), *Virtual Water Trade: Proceedings of the International Expert Meeting on Virtual Water Trade.* Value of Water Research Report Series #12.

Parveen, S. and Faisal, I. M. (n.d.). Trading Virtual Water Between India and Bangladesh: A Politico-Economic Dilemma. Retrieved from http://www/siwi.org/waterweek

Rockström, J. and Gordon, L. (2001). Assessment of Green Water Flows to Sustain Major Biomes of the World: Implications for Future Eco Hydrological Landscape Management. *Phys. Chem. Earth(B), 26*, 843–851.

Rosegrant, M. W., Paisne, M., Meijer, S. and Witcover, J. (2001). 2020 Global Food Outlook: Trends, Alternatives and Choices: A 2020 Vision for Food, Agriculture, and Environment Initiative. Washington, DC: International Food Policy Research Institute.

Siebert, S., Portmann, F. and Döll, P. (2010). Global Patterns of Cropland Use Intensity. *Remote Sensing, 2*, 1625–1643.

Singh, O. P. (2004). Water Productivity of Milk Production in North Gujarat, Western India. Proceedings of the 2nd Asia Pacific Association of Hydrology and Water Resources (APHW) Conference (Vol. 1, pp. 442–449).

Sojamo, S., Keulertz, M., Warner, J. and Allan, A. J. (2012). Virtual Water Hegemony: The Role of Agribusiness in Global Water Governance. *Water International, 37*(2), 169–182.

Sorooshian, S., Martha, P. and Hogue, T. S. (2002). Regional and Global Hydrology and Water Resource Issues: The Role of International and National Programmes, Water Policy Article. *Aquatic Sciences, 64*(2002), 317–327.

Verma, S., Doeke, A., Kampman, D. A., Van Der Zaag, P. and Hoekstra, Y. (2008). *Going Against the Flow: A Critical Analysis of Virtual Water Trade in the Context of India's National River Linking Programme*. Value of Water Research Report Series No. 31.

Warner, J. (2003). Virtual Water – Virtual Benefits? Scarcity, Distribution, Security and Conflict Reconsidered. In A. Y. Hoekstra (Ed.), *Virtual Water Trade: Proceedings of the International Expert Meeting on Virtual Water Trade*. Value of Water Research Report Series #12.

Wichelns, D. (2001). The Role of 'Virtual Water' in Efforts to Achieve Food Security and Other National Goals, With an Example From Egypt. *Agricultural Water Management, 49*(2), 131–151.

Wichelns, D. (2003). The Role of Public Policies in Motivating Virtual Water Trade, With an Example From Egypt. In A. Y. Hoekstra (Ed.), *Virtual Water Trade: Proceedings of the International Expert Meeting on Virtual Water Trade*. Value of Water Research Report Series #12.

Wichelns, D. (2004). The Policy Relevance of Virtual Water Can Be Enhanced by Considering Comparative Advantages. *Agricultural Water Management, 66*, 49–63.

Wichelns, D. (2010). Virtual Water: A Helpful Perspective, but not a Sufficient Policy Criterion. *Water Resources Management, 24*(10), 2203–2219.

Wiesmann, D. (2006). Global Hunger Index: A Basis for Cross Country Comparison. Washington, DC: International Food Policy Research Institute.

Yang, H. (2002). Water, Environment and Food Security: A Case Study of the Haihe River Basin in China. Retrieved from http://www.rioc.org/wwf/Water_in_China_Haihe.pdf

Zehnder, A. and Armin Reller, J.B. (2002). Water Issues. Editorial. *Remote Sensing, 2*, 1625–1643. Retrieved from http://www.oekom.de/verlag/german/periodika/gaia/pdf/lese2002_04.pdf

12 Of statecraft

Managing water, energy and food for long-term national security

M. Dinesh Kumar

Managing water, energy and food for long-term national security

In today's world, when natural resources are becoming scarce, managing water, energy and food requires sound knowledge in statecraft and economics. For developing economies, the ability to maintain good economic growth rates and move people out of poverty depends heavily on how water and energy resources are managed for sustainable water supplies and energy and food security (GWP, 2011). This requires solving several complex problems relating to demand and supply of water, energy and food. Along with good technology and human capital, this more often requires hard decisions on the part of the state bureaucracies dealing with water, energy and agriculture, supported by strong political leadership (World Economic Forum, 2011).

There is a growing belief that for a democratically elected government to survive in developing economies like India, the decisions concerning the management and use of resources, which directly affect the livelihoods of the millions of ordinary people, should be driven by popular concerns rather than the macroeconomic interests, and that they have to be 'politically correct.' The underlying assumption is that hard decisions to improve economy give dividends only in the long run, and are bound to be in direct conflict with measures that can appease the common man. The latter suit the political class better than the former, as they produce immediate electoral gains. A related assumption is that the ordinary people are only interested in measures that produce immediate gains and are not concerned with long-term economic management.

To a great extent this is true for developing economies like India, where a large proportion of the people living in villages and cities are concerned with their immediate survival needs, but it is also true that there is a large chunk of the population that is quite concerned with the long-term economic growth needs. The latter are worried about the implications of the short-term populist measures such as subsidized or free electricity and free water on the growth and their own household economy. Their numbers are increasing as the economy is on the upward swing. They do not benefit from farm power and irrigation subsidies. They also do not benefit from the huge diesel subsidies. Good politics also implies that this segment of the population should not simply be ignored for the sake of a larger chunk of the population, which benefits from such populist measures. Conversely, it is also true that continuing with such populist measures may lead to crippling of the economy, forcing the governments to resort to harsher measures and leading to 'political backlash.'

The recent document of the World Economic Forum on ways to achieve global water security voices these concerns on lack of political leadership unambiguously:

> As economies grow, more of the freshwater there is left available is demanded by energy, industrial and urban systems. A massive expansion of agricultural land is one option, but this need to be undertaken in a manner that exacerbates greenhouse gas emissions, thereby amplifying the challenge of adapting to changing weather patterns. More crops from much fewer drops is another option. Yet, the agricultural sector, particularly in developing economies, have historically low levels of investment in technology and human capital as well as weak institutions. This means, it does not yet have the necessary enabling environment or the extraordinary political leadership required to deliver much, much more food and fibre, with much, much less water. (World Economic Forum, 2011, p. 24)

It is also the responsibility of the technocrats, bureaucracy and the partners in policymaking, with research-based knowledge from academics, to educate the rulers of the need to have pragmatic policies that strike a balance between 'populism' and long-term economic management, and to inform them that such an approach would be in the interest of the ruling political class. The political class in turn can educate the electorate in their constituencies. For instance, it is important for the bureaucracy to educate the political class and then the political class to convey to the peasant community that flat rate tariff for farm power only benefits the rich landlords, who account for a small percentage of the community. Further, introducing metering with 'pro rata tariff' for electricity supplied to farm sector provides a level playing ground for the small and big farmers, and that under such a mode of pricing, benefits of farm power subsidy could be distributed more equitably across the farming segments (see, for instance, Kumar et al., 2011, 2013b). This can help obtain mass support for policy reforms, and is part of statecraft.

It is the duty and responsibility of the state to frame policies, rules, administrative structures and laws from time to time to make sure that the subjects have more or less equal opportunities to access common property resources like water for beneficial uses, or that benefits from the exploitation of the natural resources like minerals are distributed equitably amongst the communities that live in the localities where they are found (Ford Foundation, 2010). It is also the duty of the state to make sure that private individuals do not exploit natural resources at public cost, and whenever that does happen, to ensure that this is duly compensated through the instruments of taxation and the like. It is the duty of the academicians and scholars to help governments design the right kind of policies, legal frameworks, institutions, pricing mechanisms and taxes on development, allocation and use of these resources based on considerations of productivity, equity, sustainability and social justice. But it is the duty of the state to make sure that the right kind of institutional regime exists to bring checks and balance for ensuring quality in academic work.

But, approaching complex economic problems with kid gloves, the academics advise the politicians and bureaucrats that providing free power, fuel subsidies, free power connections and free water would help eradicate poverty, and bring more votes, and that market-based solutions would harm the interests of the poor. They are also being advised that 'small water-harvesting systems' are viable alternatives to building large water and energy infrastructure. These would not pose social and environmental problems like large systems. This is in spite of the fact that there is hardly any evidence to that effect from any part of the developing world that such solutions can replace the more conventional water and energy systems. Such an approach is being resorted to in a haste to find a place for their ideas in political and policy circles, and to show quick policy impacts of their research. This 'politics-bureaucracy-academics' nexus is costing the economies in these countries hugely.

While the argument of inefficient and unsustainable resource use is being used by a few scholars and development thinkers to counter the government policy of the pervasive subsidy, a more virulent counterargument is that attempts to do away with subsidies in water and electricity sectors would be 'political suicide' for ruling parties. But the experience of West Bengal, Uttarakhand and Gujarat with farm power metering and pricing show that this is far from the reality. For instance, the state of West Bengal is highly agrarian and is one of the poorest states in the country. The left government there could raise power tariff without any resistance from the farming community, and the power pricing policy is now continued by the subsequent government. The hugely popular government in Gujarat, led by the present Chief Minister, had raised (connected load-based) power tariff significantly, and had also been charging for new agricultural power consumers on pro rata basis. On the contrary, governments often resort to such 'sops' in order to survive the revolt, from within ruling party or coalition or from other parties across the political spectrum, on issues of corruption and mal-governance. The weak governments seem to resort to such political bribing to stay in power.

As regards small water-harvesting alternatives to large water systems, they are being opted by the governments largely as an escape route from the hassle of overcoming the fierce opposition to large water and energy systems by local and international environmental groups. On the one hand, it is hardly recognized that unwillingness to charge for water and energy from their major consumers only increases the demand for these resources, thereby increasing the need for building large systems to meet future deficits. On the other hand, the 'ostrich-like' approach of deferring and delaying investments on large water- and power-generation projects only increases the future need to either go for such large systems or increase water and energy tariffs drastically to save the economy from total collapse.

The purpose of this book is to trigger an informed debate on some of the most controversial and yet unresolved issues concerning water-energy-food security nexus in developing countries. The objective is to help politicians and bureaucrats with the right analyses and evidence to make policy choices that would contribute toward long-term water, energy and food security. In the process, the book challenged some of the popular notions about the potential impacts of some of the widely talked about policy choices in the water and energy sector in the recent times, with the support of the empirical work, which used robust methodological and analytical frameworks.

Chapter 1 discussed the manner in which the nexus between water systems, energy systems and agricultural production systems get played out at various levels – from global to regional to local – in developing countries located in the hot and arid tropics; and how framing and implementation of policies relating to supply, regulation of pricing of water and energy resources, which are not informed by sound knowledge of this nexus, could affect water, energy and food securities in those countries, using experience from India and from across the world.

Chapter 2 provided strong empirical evidence for the argument espoused by Kumar et al. (2008) and Kumar (2009) that large water systems are essential for sustaining agricultural growth in semi-arid regions. The chapter, which critically examined the 'miracle growth' of agriculture in recent years in Gujarat, showed that this 'growth' was nothing but a good recovery from a major dip in production which occurred during the drought years of 1999 and 2000 because of 4 consecutive years of successful monsoon and bulk water transfer through the Sardar Sarovar project. The 'real growth' in Gujarat's agriculture occurred during the period from 1988 to 1998, when the state did not witness any droughts. While reiterating the critical role of large water systems in regions with high spatial imbalance in resource availability and demand, it also raises serious questions about the methodology being followed for estimating agricultural growth in regions that are characterized by climates with high variability in successive years.

Chapter 3 critically evaluated the effectiveness of NREGA vis-à-vis its generating rural employment and ensuring local water security. The chapter also reviewed the design of the scheme vis-à-vis job entitlement, in terms of its potential to generate jobs and impact on labor dynamics in different situations. It argued that the nature of water-management activities chosen under the scheme and the callous way in which these activities are planned and implemented in different regions, without any consideration to their physical and socio-economic realities of the regions concerned, are creating several negative welfare effects.

It also showed that many regions do not offer sufficient scope for public works, which can generate gainful employment for the unemployed, with the result that it did not stop the out-migration. However, in many other regions, the demand for wage labor was so little that the public funds available for payment of wages are being siphoned out through 'manufactured' works. In other areas, the large-scale absorption of wage laborers in public works resulted in a shortage of laborers to work in farms, with resultant distortion in labor market. The authors identified three broad and distinct regional typologies in India for deciding the nature of water-management interventions for each region, and proposed the types for water-management works under NREGS for each typology, which has the potential to generate labor demand, while producing welfare effects.

Chapter 4 showed that in the case of hydropower projects, developing a proper mechanism to share the project benefits with the affected local population and communicating the same to take them is part of a larger confidence building measure, which is crucial to ensure the success of such developmental projects. It could go a long way in eliciting the participation of local communities in such developmental efforts. Further, compliance with the environmental safeguards, public hearings and public consultations based on participatory approach should be an integral part of the project implementation process.

Chapter 5 empirically showed that irrigation – expressed in terms of irrigated area per capita population – has a positive impact on rural poverty reduction, and the impacts are sharper in areas where the aggregate irrigated area is large. Further, the poverty reduction capacity of irrigation is found to weaken over time.

Analysis presented in Chapter 6, using empirical evidence from farmers in eastern UP, south Bihar and north Gujarat on crop-level and farm-level water use, water productivity in physical and economic terms, and groundwater pumping rates per unit land area, showed that raising power tariffs in the farm sector to achieve efficiency and sustainability of groundwater use is both socially and economically viable. With millions of farmers scattered all across the remote rural areas, it showed how the technological advancements such as prepaid electronic meters, which are operated through satellite and Internet technology, could be used to reduce the transaction cost of metering electricity use and restrict farmers from pilfering electricity. It also discussed five different options for power supply, metering and energy pricing in the farm sector and the expected outcomes of implementing each vis-à-vis efficiency of groundwater and energy use, equity in access and sustainability of groundwater. It concluded that establishing an energy quota for each farm based on sustainability considerations, and metering and charging pro rata for power used are the best options to manage groundwater and the energy economy.

Chapter 7 showed that the rising cost of diesel increased the cost of well irrigation for diesel well owners to an extent of 32% in south Bihar and 18% in eastern UP from 1990–2006. But this had no impact on the price at which water is sold by diesel well owners due to the reducing monopoly power of diesel well owners over time. The actual price at which water is available to the water buyers came down by 38% in south Bihar and 7.5% in eastern UP. Comparative analysis of irrigation water use, income from crops, dairying and entire farms, as well as water productivity in crop and milk production and at the farm level of three different categories of farmers, viz., water buyers in electric well commands, diesel well owners and water buyers in diesel well commands showed that higher cost of irrigation water motivated farmers to use irrigation water more efficiently from a physical point of view to minimize the cost of irrigation. Further, the farmers who are paying higher cost for irrigation water use it more efficiently also from agronomic and economic points of view. The net income return farmers obtain from irrigated farming is not found to be elastic to the cost of irrigation water.

Overall, the impact of diesel price hike on irrigation cost incurred by diesel well owners was not found to be significant. Also, this burden could not be passed on to the water buyers, owing to the lowering monopoly power of pump owners. Further, the study found that farmers would be able to cope with a steep rise in irrigation costs through irrigation efficiency improvements and allocating more area under crops that give higher returns per unit of land and water. This meant that rises in cost of diesel in real terms cannot make any negative impacts on economic prospects of diesel well irrigators, including water buyers. However, this does not mean that the government could keep raising diesel prices. Due consideration has to be given to the farming system resilience, particularly the kind of crops that would be economically viable for the farmers at the new input cost regimes, and the risks (both crop and market) involved.

Chapter 8 analyzed the impact of the recent policy decision of the West Bengal government to offer heavily subsidized power connections for well irrigation, and to remove the restrictions on issuing permits for drilling new energized wells on West Bengal's agriculture, energy economy and water ecosystems. The focus particularly was on the equity impacts vis-à-vis access to groundwater for irrigation for different socio-economic segments, by assessing how the pump irrigation markets would be influenced by the policy shift. The analysis presented in this chapter drives home the fact that, while these policies would do no good to the agriculture of West Bengal, they would surely and certainly result in long-term damage to the state's water and energy economy. The authors argued that implementation of the new policy would only lead to a windfall gain for the existing diesel pump owners, as they would be able to produce water at a cheap rate and sell it to poor farmers at prohibitive prices.

A policy intervention that is based on a strategy for intensifying the use of land and water will not work in a region that already has very high land-use intensity. A new policy for agricultural growth, which is driven by the strategy of enhancing the productivity of land and water and which is built on the concept of multiple-use systems, needs to be adopted in the region. Since eastern India's agro-ecology is not homogeneous, the technical intervention for multiple-use systems needs to be designed keeping in view the opportunities and the constraints of the locality.

While the official agencies concerned with rural water supplies in Indian states are preoccupied with outdated norms of 40 lpcd as the 'per capita water supply' for rural households, our research in Maharashtra shows that this approach is far less than adequate even to meet the domestic water requirements. Water is also needed for meeting requirements of kitchen gardening, livestock feeding and cottage industries, particularly for poor rural households that do not own farmland and irrigation facilities for livelihood enhancement. Inadequate water supplies can lead to several problems related to health, hygiene and nutrition for the family members, or increase the hardship of family members, especially women who are engaged in collecting water for household needs. The conventional approach of water supply surveillance won't be adequate to identify the households that actually suffer from insufficient water supplies.

Chapter 9 constructed a composite index that can assess the vulnerability of rural households to problems associated with lack of water for domestic and productive needs. The index captures the factors that influence the water access and use of the households, the socio-economic profile of the households, the physical settings of the localities/regions under consideration such as drought proneness and climate, the institutional environment, and hence has five sub-indices. The number of 'minor' factors that were considered to have an influence on the measure of these sub-indices, the underlying assumptions, the methods and procedure to compute and the data sources were also discussed. The maximum value assigned to the index was 1.0, indicating zero vulnerability. The minimum value for the index is zero, indicating extreme vulnerability.

The index was computed for 100 sample rural households, each in three villages of Maharashtra, representing three distinct agro-ecological and socio-economic environments. The villages were Varoshi in Satara district of western Maharashtra, Kerkatta in Latur district of Marathwada and Chikhali in Chandrapur district of Vidarbha. Subsequently, the vulnerable households, where the improvements in water supply or infrastructure are required to reduce hardship, were identified. Amongst the three villages, Kerkatta was found to have the highest number of vulnerable households, wherein the value of the index fell below 0.50. The participatory research conducted in the pilot villages confirmed that the parameters considered for assessing vulnerability were realistic. The computed index helps target the localities and households that need to be prioritized for interventions that are aimed at augmenting the existing domestic water supplies or improving access to the supply sources.

Chapter 10 analyzed how techno-institutional models chosen for rural water supply and institutional and policy framework governing the same influenced the performance of water supply schemes, which is in direct competition with water used for irrigated crop production. It analyzed comparative performance of different techno-institutional models for rural water supply schemes in the state. One major conclusion that emerged from the study was that the policy regarding selection of technology for water supply should be driven by resource-sustainability and source-sustainability considerations rather than consideration of decentralized management of the water supply scheme. Ideally, the sources, from which the physical allocation of water for domestic supply is not technically feasible (schemes based on groundwater) and whose sustainability is likely to be threatened by competition from other sectors, especially agriculture, need to give way to reservoir-based schemes from which such allocation of water is possible for dependability and source sustainability. Reservoir-based schemes are also better than groundwater-based schemes from the point of view of governance and management.

As regards institutions for managing water supply schemes, the analysis presented in the chapter argued that the nature of technology chosen for water supply based on the consideration of sustainability of the resource and source should guide the institutional choice. Since local groundwater-based rural water supply schemes won't be able to provide reliable water supplies in most situations, the thrust should be on reservoir-based schemes. Most of the large and medium reservoirs, which ensure dependable supplies, are located in a few water-abundant river basins. Hence, large-scale water transport would be required in taking it to water-scarce regions. Such schemes can offer economies of scale, only when large numbers of villages are covered under a single scheme. This would obviously offer a lot of challenges from the point of view of hydraulic engineering in design and operation. Professional centralized institutions with required technical know-how, expertise and financial resources are required to design, build and run such technically sophisticated schemes. More importantly, the institution, which designs the scheme, should ideally run it, for efficient management.

Chapter 11 analyzed the agricultural commodity trade across the globe to identify the emerging patterns, their determinants and their implications for global food and water security. Particularly, it examined whether the claim made regarding the growing foreign direct investment on land in Africa by the emerging Asian and Arab economies and their increased competition over the sources of global food supplies is a strategy to challenge the Western hegemony over 'international virtual water flow.' The chapter also examined whether the ongoing power shift in the global food market from the industrialized West to the industrializing East would provide any new governance architecture for water security and stewardship of water resources.

Involving reviews of international literature on virtual water trade, global land-use changes and emerging patterns in global agri-business, the analysis found that with globalization of economy, and with the pressure on many developed and some of the emerging economies to free part of their cropland for ecological uses, these countries are likely to witness land-use zoning decisions causing major land-use changes. Under these circumstances, the agri-business corporations from the developed countries would turn to regions that have large amounts of arable land lying uncultivated, to gain control over international food trade. In addition, those from oil-rich countries such as Saudi Arabia, with incentives from their national governments, might begin purchasing vast tracts of land in these countries not only to establish hegemony over international food trade, but also to ensure their own country's food security. On the one hand, the agricultural intensification and consequent expansion of cultivated area under corporate farming would severely affect the ability of the native communities in these regions to own farmland, thereby affecting their crop outputs and farm incomes. On the other hand, control over factors of production such as land and water would give monopoly power to the corporations in deciding food prices. The combined effect would be an adverse impact on regional food security and livelihood security of farmers.

Finally, we have to manage the demand for water for ensuring sustainable water supply for all uses, food and energy security in semi-arid and arid regions, where water endowment is poor. Given the inextricable link between water, energy and agriculture in groundwater-irrigated semi-arid and arid regions, the demand for both water and electricity for agricultural production can be managed through efficient pricing of electricity. This would help prevent groundwater overexploitation and reduce carbon emissions. Parallel to this, we also have to manage energy resources to secure water supplies for sustainable food production in regions where water resources are abundant, but energy to secure water is either in short supply or is prohibitively expensive. According to a UN report, nearly 1.5 billion people in the developing world lack access to electricity, and more than three billion people depend on biomass for heating and cooking (United Nations, 2010). There is a huge scope for increasing power generation from hydropower projects in India and other parts of Asia. It is estimated that nearly 75% of the 170 GW of hydropower being built across the world is in Asian countries (International Energy Agency, 2009).

While building large water storage facilities for power generation is going to pose new challenges from social and environmental points of view – with growing resistance to dam building from the local population against displacement and environmental groups against ecological damages – (Shah and Kumar, 2008), sharing benefits from such large water and energy projects with the affected local communities is going to be crucial in striking a balance between environment and development. Along with augmenting power-generation capacity, the scope for improving energy-use efficiency through proper pricing, of both diesel and electricity cannot be overstated, particularly in the context of developing economies.

But food security is not merely linked to water and energy security alone. Access to arable land is going to be a major determinant of food production, an important dimension of food security. This is because pressure on arable land is increasing at a much faster rate than the rate at which the population is growing, as a result of increased demand for biomass to meet clean energy needs. Due to policy incentives designed to reduce vehicular emissions, the International Energy Agency predicts that by 2030, at least 5% of the global road transport would be fuelled by biofuel. Producing these fuels – estimated to be more than 3.2 million barrels per day (International Energy Agency, 2008) – would create unsustainable tradeoffs in terms of land use (Earth Policy Institute, 2011; IWMI, 2007).

Arable land is increasingly becoming scarce in many parts of the world. Many regions in India and around the world are food insecure, in spite of being water abundant. They need to depend on food imports (Kumar and Singh, 2005). As shown by Kumar and Singh (2005), access to arable land is a major determinant of global agricultural commodity trade. With globalization of economy and with the pressure to free part of their cropland for ecological uses, many developed and emerging economies are likely to experience major land-use changes. It is quite likely that the agri-business corporations from the developed countries will turn to regions that have large amounts of arable land lying uncultivated to achieve food self-sufficiency along with gaining control over international food trade (Kumar, 2012). Unable to rely on trade to ensure their food security, fast-growing economies that need to ensure food supplies are striking land-lease deals with poorer nations that have fertile land. Japan now has three times more land abroad than at home. Saudi Arabia, Kuwait, China and South Korea have secured land deals in Sudan, Pakistan, Ethiopia and DRC (Vidal, 2010). The impact of this growing global phenomenon on regional food security and livelihood security of the native farmers needs to be understood carefully, as the agri-business corporations gain monopoly power over land and water resources in these poor countries.

References

Earth Policy Institute (2011). Data Center: Climate, Energy, and Transportation. Retrieved from http://www.earthpolicy.org/data_center/C23

Ford Foundation (2010). *Expanding Community Rights Over Natural Resources: Initiative Overview.* New York: Ford Foundation.

Global Water Partnership (2011). *The Water, Energy and Food Security Nexus: Understanding the Nexus.* Contribution for the Conference on the Water, Energy and Food Security Nexus: Solutions for the Green Economy, November 16–18, Bonn, Germany.

International Energy Agency (2008). *From 1st to 2nd Generation Bio-Fuel Technologies.* Paris: International Energy Agency.

International Energy Agency (2009). *World Energy Outlook, 2009.* Paris: International Energy Agency.

International Water Management Institute (2007). *Water for Food; Water for Life: Comprehensive Assessment of Water Management in Agriculture.* London: CABI Publishing.

Kumar, M. D. (2009). *Water Management in India: What Works, What Doesn't.* New Delhi, India: Gyan Publishing House.

Kumar, M. D. (2012). Does Corporate Agribusiness Have a Positive Role in Global Food and Water Security? *Water International, 37*(3), 341–345.

Kumar, M. D., Patel, A., Ravindranath, R. and Singh, O. P. (2008). Chasing a Mirage: Water Harvesting and Artificial Recharge in Naturally Water-scarce regions of India. *Economic and Political Weekly*, 43(35): 61–71.

Kumar, M. D., Scott, C. and Singh, O. P. (2011). Inducing the Shift From Flat-Rate or Free Agricultural Power to Metered Supply: Implications for Groundwater Depletion and Power Sector Viability in India. *Journal of Hydrology, 409*(1–2), 382–394.

Kumar, M. D., Scott, C.A. and Singh, O. P. (2013b, January). Can India Raise Agricultural Productivity While Reducing Groundwater and Energy Use? *International Journal of Water Resources Development*(online), 1–17.

Kumar, M. D. and Singh, O. P. (2005). Virtual Water in Global Food and Water Policy Making: Is There a Need for Rethinking? *Water Resources Management, 19*(6), 759–789.

Kumar, M. D., Sivamohan, M. V. K. and Bassi, N. (2013a). *Water Management, Food Security and Sustainable Agriculture in Developing Economies.* Oxon: United Kingdom: Routledge.

Shah, Z. and Kumar, M. D. (2008). In the Midst of the Large Dam Controversy: Objectives, Criteria for Assessing Large Water Storages in Developing World. *Water Resources Management, 22*, 1799–1824.

United Nations (2010). *Energy for a Sustainable Future, The Secretary General's Advisory Group on Energy and Climate Change.* New York: United Nations.

Vidal, J. (2010, March 7). How Food and Water Are Driving a 21st Century African Land Grab. *Observer.* Retrieved from http://www.theguardian.com/environment/2010/mar/07/food-water-africa-land-grab

World Economic Forum (2011). *Water Security: The Water-Food-Energy-Climate Nexus.* The World Economic Forum Water Initiative. Washington/Covelo/London: Island Press.

Index

For Product Safety Concerns and Information please contact our EU representative GPSR@taylorandfrancis.com Taylor & Francis Verlag GmbH, Kaufingerstraße 24, 80331 München, Germany